Lecture Notes in Mathematics

Edited by A. Dold and B. Eckmann

1166

Banach Spaces

Proceedings of the Missouri Conference
held in Columbia, USA, June 24–29, 1984

Edited by N. Kalton and E. Saab

Springer-Verlag
Berlin Heidelberg New York Tokyo

Editors

Nigel J. Kalton
Elias Saab
Department of Mathematics, University of Missouri
Columbia, MO 65211, USA

Mathematics Subject Classification (1980): 46

ISBN 3-540-16051-5 Springer-Verlag Berlin Heidelberg New York Tokyo
ISBN 0-387-16051-5 Springer-Verlag New York Heidelberg Berlin Tokyo

Printing and binding: Beltz Offsetdruck, Hemsbach/Bergstr.
2146/3140-543210

FOREWORD

From 24 June to 29 June, 1984, the University of Missouri hosted an NSF-CBMS Regional Conference on "Factorization of linear operators and goemetry of Banach spaces". The principal speaker was G. Pisier (Paris). There were also five invited addresses given by J. Bourgain (Brussels), D. Burkholder (Illinois), J. Lindenstrauss (Jerusalem), H. Rosenthal (Texas), M. Talagrand (Paris) and L. Tzafriri (Jerusalem). The conference was attended by 168 mathematicians from many different countries. In addition to the main speakers there were over fifty contributed talks.

Professor Pisier's lectures will be published separately in the CBMS Regional Conference Series, published by the American Mathematical Society. In this volume are included twenty-two papers contributed by participants.

We would like to acknowledge everybody who helped to make this conference a success. First, we would like to thank all the main speakers; in particular, special thanks are due to Gilles Pisier whose superb lectures formed the centerpiece of the conference.

From the point of view of organization, we gratefully acknowledge the generous financial support of the National Science Foundation and the Argonne Universities Association Trust Fund. Milton Glick, Dean of the College of Arts and Sciences (UMC), provided invaluable assistance both by helping to locate funding sources and by smoothing out some of the obstacles in our path; his constant support was much appreciated. The efficient and friendly assistance provided by Ms. Freddy Randolph and her colleagues at the UMC Office of Conferences is also gratefully acknowledged.

Finally we would like to thank all of our colleagues and friends who donated their time and energy to help in this endeavor. A partial list includes Dennis Sentilles, Paulette Saab, Peter Casazza, Janet Tremain, Dale Bachman, Al Dixon, Dean Allison, Matt Mayfield, Renee Sentilles, Musbah Abedessalam and Carolyn Eoff.

Nigel Kalton
Proceedings Editor

Elias Saab
Proceedings Editor
and Conference Director

Columbia, Missouri

TABLE OF CONTENTS

WEAKLEY CONTINUOUS FUNCTIONS ON BANACH SPACES CONTAINING ℓ_1

R. M. Aron, J. Diestel and A. K. Rajappa
Kent State University
Kent, Ohio 44242

Walsh College
North Canton, Ohio 44720

In this paper, we give a characterization of Banach spaces E containing ℓ_1 in terms of certain classes of weakly continuous functions from E to any Banach space F.

Let A be a subset of E. A function $f : A \to F$ is said to be weakly continuous if for each $x \in A$ and $\epsilon > 0$, there are $\phi_1, \phi_2, \ldots, \phi_n$ in E^* and $\delta > 0$ such that if $y \in A$, $|\phi_i(x-y)| < \delta$ for all $i = 1, 2, \ldots, n$, then $\|f(x) - f(y)\| < \varepsilon$. We denote $C_{wb}(E,F)$ (respectively $C_{wk}(E,F)$) the space of all functions from E to F which are weakly continuous when restricted to bounded sets (respectively weakly compact sets). Clearly we have $C_{wb}(E,F) \subset C_{wk}(E,F)$. We say a function $f : E \to F$ is weakly sequentially continuous if it takes weakly convergent sequences into norm convergent ones. We denote $C_{wsc}(E,F)$ the class of all weakly sequentially continuous functions from E to F. Ferrara, Gil and Llavona [3] have shown that if E is a Banach space that contains no subspace isomorphic to ℓ_1, then for every Banach space F, $C_{wb}(E,F)=C_{wk}(E,F)$. They raised the question: -
If E is a Banach space that contains ℓ_1, is it always true that $C_{wb}(E,F) \subsetneqq C_{wk}(E,F)$ for every Banach space F?
We show here that the affirmative answer to this question is an easy consequence of facts about absolutely 2-summing operators between Banach spaces.

Theorem:

A Banach space E contains an isomorphic copy of ℓ_1 if and only if $C_{wb}(E,F) \subsetneqq C_{wsc}(E,F)$ for all Banach spaces F.

Proof:

From observations made in [3] if E does not contain an isomorphic copy of ℓ_1 then $C_{wb}(E,F) = C_{wsc}(E,F)$ for all Banach spaces F.

The proof of the other part relies on Pelczynski's observation that if E

contains an isomorphic copy of ℓ_1, then there exists a linear operator $S : E \to \ell_2$ such that S is absolutely 2-summing. To be sure, the quotient map q from ℓ_1 onto ℓ_2 is absolutely summing and hence absolutely 2-summing. Therefore q admits a factorization in the form

$$\begin{array}{ccc} \ell_1 & \xrightarrow{\quad q \quad} & \ell_2 \\ {\scriptstyle A}\downarrow & \# & \uparrow{\scriptstyle B} \\ L_\infty(\mu) & \xrightarrow{\quad I \quad} & L_2(\mu) \end{array}$$

where μ is a regular Borel probability measure on some compact Hausdorff space, $I : L_\infty(\mu) \to L_2(\mu)$ is the natural inclusion and $A : \ell_1 \to L_\infty(\mu)$ and $B : L_2(\mu) \to \ell_2$ are bounded linear operators; all this we get thanks to the Grothendieck-Pietsch Domination theorem ([2], p. 60). Of course, A extends to a bounded linear operator G from E to $L_\infty(\mu)$ by $L_\infty(\mu)$'s injectivity; I is absolutely 2-summing and so $BIG = S : E \to \ell_2$ is an absolutely 2-summing quotient map of E onto ℓ_2. Further S being a quotient map onto ℓ_2, S is noncompact.

Now we are ready to complete the proof of necessity in the theorem. An absolutely 2-summing operator sends weakly convergent sequences into norm convergent ones ([2], p. 61). Therefore the function $\|\cdot\| \circ S \in C_{wsc}(E,\mathbb{R})$. We show this function $\|\cdot\| \circ S \notin C_{wb}(E,\mathbb{R})$. S being noncompact, S is not weakly continuous on bounded sets [1]. This implies that there exists an $\varepsilon>0$ and a net (x_α) converging weakly to x with x, $x_\alpha \in$ unit ball $B_1(E)$ such that $\|Sx_\alpha - Sx\| > \varepsilon$. This, in turn, means that there exists an $\varepsilon > 0$ and a net $(x_\alpha - x)$ converging weakly to 0 with $x_\alpha, x \in B_1(E)$ such that $(\|\cdot\| \circ S)(x_\alpha - x) > \varepsilon$. Hence $(\|\cdot\| \circ S)(x_\alpha - x)$ is not convergent to zero. Thus $\|\cdot\| \circ S \notin C_{wb}(E,\mathbb{R})$. Hence $C_{wb}(E,\mathbb{R}) \subsetneq C_{wsc}(E,\mathbb{R})$ if E contains ℓ_1.

Finally we want to show $C_{wb}(E,F) \subsetneq C_{wsc}(E,F)$ for all Banach spaces F when E contains ℓ_1. As we have shown $C_{wb}(E,\mathbb{R}) \subsetneq C_{wsc}(E,\mathbb{R})$ when E contains ℓ_1, choose an $f \in C_{wsc}(E,\mathbb{R})$ but not in $C_{wb}(E,\mathbb{R})$

Let $y \in F$ with $\|y\| = 1$. Consider the function $g : E \to F$ given by $g(e) = f(e)y$ for $e \in E$. As $f \in C_{wsc}(E,\mathbb{R})$, $g \in C_{wsc}(E,F)$. If we suppose that $C_{wb}(E,F) = C_{wsc}(E,F)$, then $g \in C_{wb}(E,F)$. Thus for $\varepsilon > 0$, $x \in B_1(E)$ there exist $(\phi_i)_{i=1}^n \subset E^*$ and $\delta > 0$ such that for all $x^1 \in B_1(E)$ with $|\phi_i(x^1 - x)| < \delta$ for every $i = 1, \ldots, n$, we have $\|g(x^1) - g(x)\| < \varepsilon$. That is $|f(x^1) - f(x)|\ \|y\| < \varepsilon$, and so $|f(x^1) - f(x)| < \varepsilon$. Hence $f \in C_{wb}(E, \mathbb{R})$, a contradiction to the choice of f.

REFERENCES

[1] R. M. Aron and J. B. Prolla, Polynomial approximation of differentiable functions on Banach spaces, J. Reine-Angew Math (Crelle), 313 (1980) p. 195-216.

[2] J. Diestel, Sequences and series in Banach spaces, Springer-Verlag, Graduate Texts in Mathematics, 1984.

[3] J. Ferrera, J. Gomez Gil and J. G. Llavona, On completion of spaces of weakly continuous functions, Bull. London Math. Soc. 15(1983), 260-264.

SOME REMARKS ON THE BANACH SPACE STRUCTURE OF THE BALL-ALGEBRAS

J. Bourgain
University of Brussels
Pleinlaan 2-F7, 1050 Brussels
Belgium

Denote $B = \{\zeta \in \mathbb{C}^m;\ |\zeta| = <\zeta,\zeta>^{\frac{1}{2}} \leq 1\}$ the closed unit ball in $\mathbb{C}^m (m>1)$ and $S = \partial B$ the sphere equipped with normalized Haar measure σ. Let $A(B)$ stand for the ball algebra, i.e. the space of continuous functions on B which are analytic inside B. The restriction map $f \to f|_S$ identifies $A(B)$ with a subspace of $C(S)$ the Banach space of continuous functions on S. In [3], it is proved that $A(B)$ fails the so-called $(i_p - \pi_p)$ property for $p \neq 2$, emphasizing a different behaviour with respect to the disc algebra $A(D)$ (see [4] for details). This fact provides also the first proof of non-isomorphism of $H^\infty(B_m)$ for $m > 1$ and $H^\infty(D)$. The problem whether or not the dual space $A(B)^*$ has cotype 2 (or any finite cotype) or satisfies Grothendieck's theorem is unsettled at the time of writing.

If μ is a positive Radon measure on S, denote for $1 \leq p < \infty$ by $H^p(\mu)$ the closure of $A(B)$ in the space $L^p(S;\mu)$. The following phenomenon is a strong version of the non-existence of generalized Cauchy projections (cf. [2], section 1).

<u>THEOREM 1</u>: There exists $\mu \in M_+(S)$ such that $H^p(\mu)$ is not an L^p space for $1 < p < 2$.

The measure μ can be choosen in $L^1(S;\sigma)$. For instance in the case m=2, with parametrization $\zeta = (z,w)$, $z = \sqrt{\rho}\, e^{i\theta}$, $\omega = \sqrt{1-\rho}\, e^{i\psi}$, μ has the form $d\mu = \Delta(\rho, \Theta-\psi)\, d\rho\, d\theta\, d\psi$, $\Delta \in L^1_+ (d\rho\, d\eta)$.

<u>THEOREM 2</u>: For $2 < p < \infty$, there exists a (bounded, linear) operator u from $A(B)$ into ℓ^p with no linear extension to $C(S)$.

Notice that in [1] it is proved that each operator from $A(D)$ into a Banach space of cotype p is q-integral for $q > p$.

In [3], the spaces of homogeneous polynomials of degree N, say P_N play an important role. They have the remarkable property that the orthogonal projection on $\mathcal{P}_N$

$$\mathcal{P}_N\, f = \mathcal{P}\, f = C_N^{-1} \int \langle \eta, \zeta \rangle^N\, f(\zeta)\, \sigma(d\zeta) \quad , \quad C_N = \int |\zeta_1|^{2N}\, d\sigma$$

are bounded under the L^p-norm $(1 \le p \le \infty)$. Moreover, it is important that for $f \in A(B)$, $\mathcal{P}_N f$ is also expressed as

$$\mathcal{P}_N f(\zeta) = \frac{1}{2\pi} \int_0^{2\pi} f(\zeta\, e^{i\theta})\, e^{-iN\theta} d\theta \ .$$

The key lemma is the following (see [3], lemma 3).

<u>LEMMA 1</u>: For $\varepsilon > 0$, $n = 1, 2, \ldots$ fixed, there are 1-bounded homogeneous polynomials $\{p_j\}_{j=1}^n$ on B, $d(p_j) = \text{degree}(p_j) = N_j$, for which the sets $[\zeta \in S\ ; |p_j(\zeta)| > \varepsilon]$ are disjoint but $\| \sum_{1 \le j \le n} |q_j| \|_\infty > c\, n\, \varepsilon$ whenever $\{q_j\}_{j=1}^n$ is a sequence in $A(B)$ satisfying $\|p_j - q_j\|_\infty < c\ (1 \le j \le n)$.

Here $c > 0$ is numerical and $\{N_j\}$ any sequence provided it increases rapidly enough.

Both Theorems 1 and 2 appear as reinterpretations of Lemma 1.

Let us point the following fact out:

<u>LEMMA 2</u>: For $\{N_j\}$ sufficiently increasing and $\alpha_j \in \mathcal{P}_{N_j}$,

$$(1) \qquad \|\Sigma\, \alpha_j\|_{A(B)^*} \sim \sup \{\Sigma\, |\langle \alpha_j, \psi_j \rangle| ;\ \psi_j \in \mathcal{P}_{N_j} \ ,\ \Sigma |\psi_j|^2 \le 1\}$$

holds. Hence $\{\mathcal{P}_{N_j}\}$ is an unconditional decomposition of $[\mathcal{P}_{N_J} ; j=1,2,\ldots]$ considered as subspace of $A(B)^*$ and admits a lower 2-estimation (up to numerical constant).

<u>Proof</u>: The second statement follows clearly from the first. Since

$$\|(\Sigma\, |\mathcal{P}_N\, f|^2)^{\frac{1}{2}}\|_\infty \le \|f\|_\infty \quad \text{for} \quad f \in A(B)$$

we have $||\Sigma\, \alpha_j||_* \le \sup \Sigma| < \alpha_j, \psi_j >|$. For the reverse inequality, fix $\{\psi_j\}_{j=1}^{J}$, $\psi_j \in P_{N_j}$, $\Sigma|\psi_j|^2 \le 1$ and use a Rudin-Shapiro type construction in the following way. Define inductively

$$\begin{cases} f_1 = \psi_1 \\ g_1 = 1 \end{cases} \qquad \begin{cases} f_{j+1} = f_j + K\,[\bar{g}_j\, \psi_{j+1}] \\ g_{j+1} = g_j - \bar{f}_j\, \psi_{j+1} \end{cases}$$

where K is the Cauchy projection.

For N_{j+1} large enough with respect to $N_1,\dots,N_j$, it follows that $K[\bar{g}_j\, \psi_{j+1}] \approx \bar{g}_j\, \psi_{j+1}$ (by elementary properties of Toeplitz operators). Hence

$$|f_{j+1}|^2 + |g_{j+1}|^2 \approx (1 + |\psi_{j+1}|^2)\ (|f_j|^2 + |g_j|^2)\ .$$

Thus $f_J \in A(B)$, $||f_J|| \le C$ and

$$|< \Sigma\, \alpha_j\ , f_J >| \approx \Sigma < \alpha_j\ , \bar{g}_{j-1}\, \psi_j > = \Sigma < \alpha_j\ , \psi_j > .$$

Proof of Theorem 1: It suffices for fixed $1 < p < 2$ to exhibit $H^p(\mu)$-spaces on the ball which do not admit uniformly complemented embedding in an $L^p(\nu)$-space. The procedure to glue together is straightforward since $A(B)$ is known to be isomorphic to its c_o-direct sum (P. Wojtaszczyk, 1983 preprint, [5]).

Fix an integer n, $p < r < 2$ and $\varepsilon = n^{-1/r}$. Application of Lemma 1 and and a separation argument provides homogeneous polynomials $\{p_j\}_{j=1}^{n}$ and a Radon probability measure μ on S satisfying

(1) $||p_j||_\infty \le 1$, $[|p_j| > n^{-1/r}]$ disjoint

(2) $\int \max |\psi_j - p_j|\, d\mu \ge c$ or $\Sigma \int |\psi_j|\, d\mu \ge c\, n^{1/r'}$ if $\{\psi_j\} \subset A(B)$

Suppose now T a linear isomorphism from $H^p(\mu)$ into an $L^p(\nu)$-space and

P a projection onto the range of T. By hypothesis, for $\{\psi_j\} \subset H^p(\mu)$

(3) $$\|(\Sigma |\psi_j - p_j|^2)^{\frac{1}{2}}\|_{L^p(\mu)} \geq c \quad \text{or} \quad (\Sigma \|\psi_j\|^p_{L^p(\mu)})^{1/p} \geq c\, n^{1/p - 1/r}$$

By the square function property and since T is an isomorphism, (3) implies that for $\{y_j\} \subset L^p(\nu)$, taking $x_j = Tp_j$

(4) $$\|(\Sigma |x_j - y_j|^2)^{\frac{1}{2}}\|_{L^p(\nu)} \geq c\, \|P\|^{-1}\, \|T^{-1}\|^{-1}$$

or

$$(\Sigma \|y_j\|^p)^{1/p} \geq c n^{1/p - 1/r}\, \|P\|^{-1}\, \|T^{-1}\|^{-1}$$

hold.

Denoting $\lambda_j = \mu[|p_j| > n^{-1/r}]$, $\Sigma\, \lambda_j \leq 1$ and clearly

(5) $$\|(\Sigma |a_j|^2\, |x_j|^2)^{\frac{1}{2}}\|_p \leq \|T\| (\Sigma |a_j|^p \lambda_j)^{1/p} + n^{-1/r} \|T\|\, (\Sigma |a_j|^2)^{\frac{1}{2}}$$

Let $\delta > 0$ be a number to be made precise later and $y_j = x_j\, \chi_j$ where χ_j is the indicator function of the set $A_j = [|x_j| > \delta\, (\Sigma |x_k|^2)^{\frac{1}{2}}]$. Notice that $\|\Sigma\, \chi_j\|_\infty \leq \delta^{-2}$.

First, using the Schwartz inequality and (5), it follows

$$\begin{aligned} \|(\Sigma |x_j - y_j|^2)^{\frac{1}{2}}\|_p &\leq \|\max |x_j - y_j|^{\frac{1}{2}}\, (\Sigma (|x_j - y_j|)^{\frac{1}{2}}\|_p \\ &\leq \delta^{\frac{1}{2}}\, \|(\Sigma |x_j|^2)^{\frac{1}{2}}\|_p^{\frac{1}{2}}\, \|\Sigma |x_j|\,\|_p^{\frac{1}{2}} \\ &\leq \delta^{\frac{1}{2}}\, \|T\|^{\frac{1}{2}}\, \|T\|^{\frac{1}{2}}\, (\Sigma\, \|p_j\|_p)^{\frac{1}{2}} \\ &\leq \delta^{\frac{1}{2}}\, \|T\|\, (\Sigma\, \lambda_j^{1/p} + n^{1/r'})^{\frac{1}{2}} \qquad (6) \\ &\leq \delta^{\frac{1}{2}}\, n^{1/2r'}\, \|T\| \end{aligned}$$

Next, letting $\{\gamma_j\}$ be independent r-stable variables and applying successively Hölder's inequality and (5), we also get

$$(\Sigma||y_j||_p^p)^{1/p} \le ||(\Sigma\ \chi_j)^{1/p-1/r}(\Sigma|x_j|^r)^{\frac{1}{r}}||_p$$

$$\le \delta^{-2(1/p-1/r)} \int ||\Sigma\ \gamma_j\ (\omega)x_j\ ||_p\ d\omega$$

$$\le \delta^{-2(1/p-1/r}\ ||T||\ \{\int \Sigma\ \lambda_j|\gamma_j|^p)^{1/p} + n^{-1/r}\int(\Sigma|\gamma_j|^2)^{\frac{1}{2}}\}$$

$$(7) \qquad \le C_{p,r}\ \delta^{-2(1/p-1/r)}\ ||T||$$

As a consequence of (4), (6), (7)

$$||P||\ \ ||T||\ \ ||T^{-1}|| \ge c \min\{(\delta^{-1}\ n^{-1/r'})^{\frac{1}{2}},\ (\delta^2\ n)^{1/p-1/r}\}$$

Choosing $\delta = n^{-\rho}$ $(\frac{1}{r'} < \rho < \frac{1}{2})$ shows that left member $\to \infty$ if $n \to \infty$.

We now pass to Theorem 2. We use the following fact.

<u>LEMMA 3</u>: Let $\{e_j\}_{j=1}^n$ be a normalized unconditional sequence in a Banach space satisfying a lower 2-estimation and the upper estimation

$$||\Sigma e_j|| < \sigma\ n \quad \text{for some } \sigma > 0\ .$$

Then there exists $A \subset \{1,\dots,n\}$, $|A| > \frac{n}{2}$ such that $||\Sigma_A\ a_i\ e_i|| \le$ $K \log n\ (\Sigma_A|a_i|^p)^{1/p}$ for all scalars $\{a_i\}_{i\in A}$ where $K \sim \sigma^{2/q}n^{1/q}$, $q = \frac{p}{p-1}$. and $p<2$.

<u>Proof</u>: Denote $\mathcal{C} = \{I \subset \{1,\dots,n\}\ ;\ ||\Sigma_I\ e_i|| > K|I|^{1/p}\}$ and let $\mathcal{M} \subset \mathcal{C}$ be a maximal subfamily of $\mathcal{C}$ consisting of disjoint sets. It will suffice to show that $\Sigma_{\mathcal{M}}|I| < \frac{n}{2}$ since, by the triangle inequality, $A = \{1,\dots,n\} \setminus \bigcup_{I\in\mathcal{M}} I$ will satisfy. Using the lower 2-estimation, we get

$$\sigma\ n \ge ||\Sigma_{\mathcal{M}}\ \Sigma_I\ e_i|| \ge K(\Sigma\ |I|^{2/p})^{\frac{1}{2}} \ge K^{q/2}\ (\Sigma_{\mathcal{M}}|I|)^{\frac{1}{2}}$$

since $I \in \mathcal{C}$ implies $|I| > K^q$. Substitution of K yields the estimate on $\Sigma_{\mathcal{M}}|I|$.

<u>Proof of Theorem 2</u>: Apply again for fixed $1 < p < 2$, n a positive integer and $\varepsilon = n^{-1/q}$, Lemma 1. Notice that the polynomials p_j fulfill

$$(1) \qquad \| (\Sigma |p_j|^q)^{1/q} \|_\infty \le 1$$

Defining

$$F = \{ (q_1, \ldots, q_n) \; ; \; q_j \in A(B) \; , \; \Sigma |q_j| \le c \, n^{1/p} \} ,$$

the property

$$\operatorname*{dist}_{\oplus_\infty C(S)} \; (\, (p_1, \ldots, p_n) \; ; \; F) > c$$

provides measures $\mu_1, \ldots, \mu_n \in M(S)$ fulfilling

$$(2) \quad \Sigma \, \| \mu_j \| \le 1 \; ; \; \Sigma |\langle p_j, \mu_j \rangle| \ge c + \sup_F \Sigma |\langle \mu_j, q_j \rangle|$$

By the special form of the orthogonal projections from $A(B)$ onto the spaces of homogeneous polynomials, $\{q_j\} \in F \Rightarrow \{P_{N_j} q_j\} \in F$.

Hence, defining $\xi_j = P_{N_j}[\mu_j]$, again

$$(3) C \ge \Sigma \| \xi_j \|_1 \ge \Sigma |\langle \xi_j, p_j \rangle| \ge c + \sup_F \Sigma |\langle \xi_j, q_j \rangle| \ge c \, n^{1/p} \max_j \| \xi_j \|_1$$

Normalizing, write $\xi_j = \lambda_j \, x_j$, $\| x_j \|_1 = 1$, noticing that $\lambda_j \le C \, n^{-1/p}$. Since $\Sigma \, \lambda_j \, |\langle x_j, p_j \rangle| \sim \Sigma \, \lambda_j$, it follows that there is a subset A of $\{1, \ldots, n\}$, $|A| = m > c \, n^{1/p}$ so that

$$|\langle x_j, p_j \rangle| > c \quad \text{and} \quad \lambda_j > m^{-1} (\log n)^{-1} \quad \text{for} \quad j \in A.$$

It follows from (3) that

$$\int \| \Sigma_A \, \varepsilon_j \, x_j \|_{A(B)^*} \, d\varepsilon \le m (\log n) \sup \{ \sum_{j \in A} | \langle \xi_j, q_j \rangle | \; ; \; q_j \in A, \; \Sigma |q_j|^2 \le 1 \}$$

$$\le m^{3/2} \, (\log m) \, n^{-1/p}$$

Apply Lemma 3 to the sequence $\{e_j = x_j \otimes \varepsilon_j\}_{j\in A}$ in Rad A(B)*, which by Lemma 2 has lower 2 estimation. We replace here n by m and $\sigma = m^{1/2-1/p}\,\ell og\, m$. This gives a subset $A_1 \subset A$, $|A_1| > \frac{m}{2}$ such that

$$(4)\qquad \int \|\Sigma_{A_1}\ \varepsilon_i\ a_i\ x_i\ \|_{A(B)^*} \leq K(\log m)\ (\Sigma\ |a_i|^p)^{1/p}$$

where
$$K \sim m^{1/q}\ \sigma^{2/q} = m^{2/q^2}\ \log m$$

Denote Δ the Can or group $\{1,-1\}^m$ and suppose v a linear extension to $C(\Delta\times S)$ of the linear map

$$A(B) \sim C_{A(B)}(\Delta) \xrightarrow{\ u\ } \ell^q_{|A_1|} : F \longrightarrow (\langle F, \varepsilon_j \otimes x_j \rangle)_{i\in A_1}\ .$$

Defining $k_j(t) = \int v^*(e_j)(\varepsilon,t)\ \varepsilon_j\ d\varepsilon$, clearly k_j and x_j induce the same element of A(B)*. Hence $|\langle k_j, p_j\rangle| > c$. Also, by (1)

$$(5)\quad c\ m \leq \Sigma|\langle k_j,p_j\rangle| \leq \int (\Sigma|k_j|^p)^{1/p} \leq \|v\| \int \|\ \Sigma\ \gamma_j\ e_j\|_{\ell^p} \leq \|v\| m^{1/p}\ \log m$$

denoting $\{\gamma_j\}_{j\in A_1}$ a sequence of independent p-stable variables.

As a consequence of (4), (5)

$$\frac{\|v\|}{\|u\|} \geq c\ (\log\ m)^{-3}\ m^{1/q-2/q^2} \quad \text{tending to } \infty \text{ if } n \to \infty.$$

This completes the proof.

REFERENCES

[1] J. Bourgain, New Banach space properties of the disc algebra and H^∞, Acta Math., Vol 152, 1984, 1-48.

[2] J. Bourgain, Bilinear forms on H^∞ and bounded bianalytic functions, Trans. AMS, Vol 287, 1985, 1-25.

[3] J. Bourgain, Applications of the spaces of homogeneous polynomials to some problems on the ball algebra, Proc. AMS, Vol 93, NL, 1985, 277-283

[4] A. Pelczynski, Banach spaces of analytic functions and absolutely summing operators, Reg. Conf. Soc. in Math. No. 30, Providence 1977.

[5] P. Wojtaszczyk, Projections and isomorphisms of the ball algebra, preprint P.A.N., to appear in Studia Math.

NUMERICAL RADIUS ATTAINING OPERATORS

Carmen Silvia Cardassi
Department of Mathematical Sciences
Kent State University
Kent, Ohio 44242

1. Introduction

We show that the set of numerical radius attaining operators from ℓ_1 into itself is dense in the space of all operators. We also establish the same result in case of operators on c_o. The proofs make essential use of the form such operators must take and are, therefore, completely elementary.

We discuss also the situation of other classes of Banach spaces, specifically, C(K)-spaces, $L_1(\mu)$-spaces and some reflexive spaces.

2. Preliminaries

We introduce initially some definitions and notations.

If X is a Banach space and $T:X \to X$ is a bounded linear operator, we define the numerical radius of T by

$$v(T) = \sup \{ \ | <x^*,Tx> | \ : \ (x,x^*) \ \varepsilon \ \pi(X) \ \} ,$$

where $\Pi(X) = \{ \ (x,x^*) \ \varepsilon \ X \times X^* : \ ||x|| = ||x^*|| = x^*(x) = 1 \ \}$. Note that $v(T) \leq ||T||$.

We say that T attains its numerical radius if there exists $(x_o,x^*_o) \varepsilon \Pi(X)$ that $v(T) = |< x^*_o,Tx_o >|$. We use e_k to denote the k-th unit vector, ie, $e_k(n) = \delta^n_k$. Also we convention that sgn a = 1 if a = 0.

L(X) will denote the space of all bounded operators from X into itself and NRA(X) the subset of L(X) consisting of the numerical radius attaining operators.

In [1] Berg and Sims proved that $\overline{NRA(X)} = L(X)$ when X is an uniformly convex space and raised the question if the same is true for any Banach space.

We will show here that $\overline{NRA(X)} = L(X)$ for $X = \ell_1$ and for $X = c_o$. We have also shown that any C(K), with K a compact Hausdorff space, any

$L_1(\mu)$ with μ a positive finite measure defined on the Borel σ-algebra of a compact Hausdorff space and any uniformly smooth space share this property. These results will appear elsewhere.

At present we do not know any example of a Banach space X for which $\overline{NRA(X)} \neq L(X)$.

For our present purposes we will make use of the following well-known representation theorems for elements of $L(\ell_1)$ and $L(c_0)$.

Theorem A. $L(\ell_1)$ can be identified with the set of infinite matrices $[A_{nk}]$ such that $\sup_k \sum_n |A_{nk}| < \infty$. The identification is given by

$T(\{a_k\}_{k\varepsilon N}) = \{\sum_k A_{nk}a_k\}_{n\varepsilon N}$ and we have $||T|| = \sup_k \sum_n |A_{nk}|$.

Theorem B. $L(c_0)$ can be identified with the set of infinite matrices $[A_{nk}]$ such that $\sup_n \sum_k |A_{nk}| < \infty$ and $\lim_n A_{nk} = 0$, $\forall k \varepsilon N$. The identification is given by $T(\{a_k\}_{k\varepsilon N}) = \{\sum_k A_{nk}a_k\}_{n\varepsilon N}$ and we have $||T|| = \sup_n \sum_k |A_{nk}|$.

3. Main Results

Lemma 1. Let $T \varepsilon L(\ell_1)$ be represented by $[A_{nk}]$. Then for every $\varepsilon > 0$ there exists $S \varepsilon L(\ell_1)$ represented by $[B_{nk}]$ such that

(a) there are $k_0, n_0 \varepsilon N$ such that $B_{nk_0} = 0$ for $n > n_0$;

(b) $\sum_{n=1}^{n_0} |B_{nk_0}| > \sum_{n=1}^{\infty} |B_{nk}| + \varepsilon$, $\forall k \varepsilon N$, $k \neq k_0$;

(c) $||S|| = \sum_{n=1}^{n_0} |B_{nk_0}|$;

(d) $||S - T|| < 3\varepsilon$.

Proof. Given $\varepsilon > 0$, choose k_0 and n_0 such that $\sum_{n=1}^{n_0} |A_{nk_0}| > ||T|| - \varepsilon$, and define

$$B_{nk} = \begin{cases} A_{nk} & \text{if } k \neq k_0 \\ A_{nk_0} + (\operatorname{sgn} A_{nk_0}) \dfrac{2\varepsilon}{n_0} & \text{if } k = k_0 \text{ and } n \leq n_o \\ 0 & \text{if } k = k_0 \text{ and } n > n_0 \end{cases}$$

Clearly $[B_{nk}]$ satisfies (a).

If $K \neq k_0$, then

$$\sum_{n=1}^{n_0} |B_{nk_0}| = \sum_{n=1}^{n_0} \left(|A_{nk_0}| + \frac{2\varepsilon}{n_0}\right) > ||T|| + \varepsilon \geq \sum_{n=1}^{\infty} |B_{nk}| + \varepsilon,$$

and $[B_{nk}]$ defines $S \in L(\ell_1)$ satisfying (b) and (c).

Also,

$$\| S - T \| = \sup_k \sum_{n=1}^{\infty} |B_{nk} - A_{nk}| = \sum_{n=1}^{n_o} \frac{2\varepsilon}{n_o} + \sum_{n=n_o+1}^{\infty} |A_{nk_o}| < 3\varepsilon ,$$

and (d) holds.

<u>Theorem 2</u>. $\overline{NRA(\ell_1)} = L(\ell_1)$.

<u>Proof</u>. Let $T \in L(\ell_1)$, represented by $[A_{nk}]$, and $\varepsilon > 0$ be given. Using Lemma 1, get $S \in L(\ell_1)$, represented by $[B_{nk}]$ with $\| S - T \| < 3\varepsilon$, $B_{nk_o} = 0$ for $n > n_o$ and $\| S \| = \sum_{n=1}^{n_o} |B_{nk_o}|$.

Define $x_o = e_{k_o} \in \ell_1$ and $x^*_o = \{(\text{sgn } B_{nk_o})(\text{sgn } B_{k_o k_o})\}_{n \in N} \in \ell_\infty = \ell_1^*$.
Clearly $(x_o, x^*_o) \in \Pi(\ell_1)$ and

$$| < x^*_o, Sx_o > | = \sum_{n=1}^{n_o} |B_{nk_o}| = \| S \| .$$

Since $v(S) \geq | < x^*_o, Sx_o > |$ and $\| S \| \geq v(S)$, it follows that $S \in NRA(\ell_1)$.

<u>Lemma 3</u>. Let $T \in L(c_o)$ be represented by $[A_{nk}]$. Then for every $\varepsilon > 0$ there exists $S \in L(c_o)$ be represented by $[B_{nk}]$ such that

(a) there are $n_o, k_o \in N$ such that $B_{n_o k} = 0$ for $k > k_o$;

(b) $\sum_{k=1}^{k_o} | B_{n_o k} | > \sum_{k=1}^{\infty} |B_{nk}| + \varepsilon$, $\forall n \in N$, $n \neq n_o$;

(c) $\| S \| = \sum_{k=1}^{k_o} | B_{n_o k} |$;

(d) $\| S - T \| < 3\varepsilon$.

<u>Proof</u>. Same procedure as in Lemma 1, with the remark that if $T \in L(c_o)$ the $\lim_n A_{nk} = 0$, $\forall k \in N$, which gives $\lim_n B_{nk} = 0$, $\forall k \in N$ and $S \in L(c_o)$.

<u>Theorem 4</u>. $\overline{NRA(c_o)} = L(c_o)$

<u>Proof</u>. Similar to the proof of theorem 2, using now Lemma 3 and $x_o = \{a_k\}_{k \in N} \in c_o$ given by

$$a_k = \begin{cases} (\text{sgn } B_{n_o k}) \ (\text{sgn } B_{n_o n_o}) & \text{if } k \leq k_o \text{ or } k = n_o \\ 0 & \text{otherwise} \end{cases}$$

and $x^*_o = e_{n_o} \in \ell_1 = c_o^*$.

References

[1] Berg, I. D. and Sims, Brailey, "Denseness of operators which attain their numerical radius", J. Austral. Math. Soc. (Series A) 36 (1984), 130-133.

The author would like to thank J. Diestel for his help.

This work was partially supported by FAPESP (São Paulo, Brasil) and was done during a visit to Kent State University. (Kent, Ohio, U.S.A).

ABSOLUTE PROJECTION CONSTANTS VIA ABSOLUTE MINIMAL PROJECTIONS

Bruce L. Chalmers
Department of Mathematics
University of California
Riverside, CA 92521

1. Introduction and Basic Theory

Let $e = (e_i,\ldots,e_n)$ be a basis for an arbitrary n-dimensional Banach space E_n with norm $\|\cdot\|$. Let $[e]$ denote the linear span of e, i.e., $[e] = \{\sum_{i=1}^{n} a_i e_i; (a_1,\ldots,a_n) \in R^n\}$; we will use the symbol $a \cdot e$ for $\sum_{i=1}^{n} a_i e_i$. Then $E_n = ([e],\|\cdot\|) \cong$ (isometrically isomorphic to) $V_n = ([v],\|\cdot\|_\infty)$ where $v = (v_1,\ldots,v_n)$ is the n-tuple of coordinate functions for a minimal set $\tilde{v}$ of points in R^n defined by $\|a\cdot e\| = \|a\cdot v\|_\infty$, $\forall\, a = (a_1,\ldots,a_n) \in R^n$, with $\|\ \|_\infty$ indicating the usual supremum norm over $\tilde{v}$. (In general $\tilde{v}$ is an (n-1)-dimensional surface in R^n.) Then the norm of a minimal projection P_{min} from $C(\tilde{v})$ onto V_n (the norm is computed from the "Lebesgue function" of P_{min}) is the absolute projection constant $\lambda(E_n)$ of E_n. Since $\|P_{min}\| = \lambda(E_n)$, we will often write $P_{min}(E_n)$ for P_{min} and refer to $P_{min}(E_n)$ as an absolute minimal projection (for E_n).

We determine P_{min} as follows. First P_{min} can be identified with an n-dimensional subspace of $C(\tilde{v})^*$, i.e. $P_{min} = [(u_1,\ldots,u_n)] = [u]$, via $P_{min}x = \sum_{i=1}^{n} (x,u_i)v_i$ where without loss $(v_i,u_j) = \delta_{ij}$, $1 \le i,j \le n$. In fact, extending our considerations from $C(\tilde{v})$ to $L^p(\tilde{v})$, $1 \le p \le \infty$ (denote $C(\tilde{v})$ by "$L^\infty(\tilde{v})$"), and employing the usual "dot product" notation $(w\cdot v)(t) = \sum_{i=1}^{n} w_i(t)v_i(t)$ for $t \in \tilde{v}$, $w\cdot v = (w\cdot v)(t)$ for all $t \in \tilde{v}$, and $|w| = \sqrt{w\cdot w}$, we have

Theorem 1 (Variational equation for P_{min})([1],[2]). Suppose $L^p(\tilde{v}) \supset X \supset V_n \subset C^1_{pcw}$, $1 \le p \le \infty$. Then $P_{min} = [u] : X \to V_n$ must satisfy ($\frac{1}{q} + \frac{1}{p} = 1$ and "$L^\infty(\tilde{v})$" = $C(\tilde{v})$)

$$\frac{1}{p} u'\cdot v \equiv \frac{1}{q} u\cdot v' \qquad (*)$$

where " ' " denotes an arbitrary 2-sided derivative along an arbitrary C^1-vector field on $\tilde{v}$.

THEOREM 2 ([1],[2]). $u_A = \dfrac{\text{ext}|A^{\frac{1}{2}}v|}{|A^{\frac{1}{2}}v|} Av$ is a fundamental solution of (*), where A is an $n \times n$ matrix of the form B^tB, we write $A^{\frac{1}{2}} = B$, and $\text{ext } x$ denotes an extremal (norming) functional for x (e.g. $\text{ext}|x| = |x|^{p-1}$ if $p < \infty$).

Note 1. The constants c, $\vec{c}$, and $\vec{c}_p$ in [1] have subsequently been determined ([2]) to be zeros; further, any linear combination of fundamental solutions is clearly also a solution of the linear homogeneous equation (*).

Example 1 ([8]). $L^1[-1,1] = X \supset V_2 = [v]$, where $v = (1,t)$. Then $P_{min} = [u_{A_1} + cu_{A_2}]$ where $A_1 = \begin{pmatrix} 1 & 0 \\ 0 & d^2 \end{pmatrix}$ and $A_2 = \begin{pmatrix} 0 & 0 \\ 0 & 1 \end{pmatrix}$ whence $A_1v = (1,t)\begin{pmatrix} 1 & 0 \\ 0 & d^2 \end{pmatrix} = (1,d^2t)$, $A_2v = (1,t)\begin{pmatrix} 0 & 0 \\ 0 & 1 \end{pmatrix} = (0,t)$, $A_1^{\frac{1}{2}}v = (1,dt)$, and $A_2^{\frac{1}{2}}v = (0,t)$. Taking $\text{ext}|A^{\frac{1}{2}}v| = 1$ for $p=1$, we have that (for certain real constants d and c)

$$P_{min} = [\frac{(1,d^2t)}{\sqrt{1 + d^2t^2}} + c\,\frac{(0,t)}{|t|}] = [\frac{(1,d^2t)}{\sqrt{1 + d^2t^2}} + (0,c\,\text{sgn}\,t)].$$

COROLLARY 1. $P_{min} = [u]$, where u is a (in general infinite) linear combination of fundamental solutions of (*). In the case $p = \infty$, which is the focus of this note, for each $t \in \tilde{v}$ we obtain fundamental solutions of the form

$$u_{A_t} = (\text{ext}|A_t^{\frac{1}{2}}v|)A_tv \qquad (1)$$

where A_t is picked to yield a maximal set <u>including</u> $A_t^{\frac{1}{2}}t$ of global maxima for $|A_t^{\frac{1}{2}}v|$ which forms the support for the measure $\text{ext}|A_t^{\frac{1}{2}}v|$. On the regions of the surface $\tilde{v}$ where the fundamental solutions u_{A_t}

vary smoothly with t, their aggregate is also given by

$$u(t) = \frac{v'_{\perp}}{w}(t) \qquad (2)$$

where $v'_{\perp}$ denotes the outward pointing normal to the $(n-1)$-dimensional surface determined by $\tilde{v}$, and w is a "weight"-function determined by the constancy on a certain portion of $\tilde{v}$ of the Lebesgue function P_{min}.

Proof: In the case $p = \infty$, $|A_t^{\frac{1}{2}}v|$ is constant on the support of the measure $\text{ext}|A_t^{\frac{1}{2}}v|$, whence the form for u_{A_t} is seen to be simply a convenient way to describe a fundamental solution, and it is easy to see that the set of global maxima can without loss be taken to be a maximal set. The (*)-equation yields $u \cdot v' \equiv 0$ from which $u(t) = v'_{\perp}(t)/w(t)$, for some scalar-valued function $w(t)$, by the definition of $v'_{\perp}$. But then also in [2] (see also [6]) it is shown that the Lebesgue function $L(x) = \int_{\tilde{v}} |\sum_{i=1}^{n} u_i(t) v_i(x)| dt$ is constant. □

Conclusion. $P_{min}(E_n) = [u]$ where u is a combination of "discrete" fundamental solutions (1) and "continuous" solutions (2).

2. Applications

Definition. E_n is a regular polyhedral space if $\tilde{v}$ can be taken to be the vertices of a regular polyhedron (e.g. if $E_n = \ell_n^1$, then $\tilde{v}$ = the corners of the ℓ_n^{∞}-ball and if $E_n = \ell_n^{\infty}$, then $\tilde{v}$ = the corners of the ℓ_n^1-ball).

THEOREM 3 ([4]). If E_n is a regular polyhedral space, then we can take $P_{min}(E_n)$ = the Fourier projection $F = \frac{1}{\|v_1\|_2^2} \sum_{i=1}^{n} v_i \otimes v_i$ where $(v_i, v_j)_2 = \delta_{ij}\|v_1\|_2^2$, $1 \le i,j \le n$; here $(f,g)_2 = \int_{\tilde{v}} fg d\mu$ where μ is the unique normalized uniformly distributed measure on $\tilde{v}$.

The proof in [4] follows by showing directly that the coordinate functions are biorthogonal and then comparing the norms of F with the absolute projection constants obtained in [7]. But the motivation for considering F is provided by Corollary 1. For $\tilde{v}$ can be taken to be just the set of vertices, in which case the set $u(t)$ of form (2) is vacuous and for each $t \in \tilde{v}$, $A_t = Id$ (identity) yields the same maximum set of global maxima of $|v|$, namely, the entire set $\tilde{v}$. Thus $P_{min} = [(\text{ext}|v|)v]$. (Also compare [3].)

Note 2. The constants $\lambda(E_n)$ for E_n in Theorem 3 are computed very easily using F ([4]). In particular, the result of [9] covering $\lambda(\ell_n^1)$ is recaptured immediately as follows.

Example 2. Let $v_i = (v_{i1}, v_{i2}, \ldots, v_{i2^n})$, $1 \le i \le n$ be the coordinate 2^n-tuples (all entries ± 1) and without loss let $v_{i1} = 1$ for all i. Let $\gamma = \dfrac{1}{\|v_1\|_2^2}$ and then by symmetry

$$\|P_{min}(\ell_n^1)\| = \sup_{\|f\|_\infty = 1} \|Ff\| = \sup_{\|f\|_\infty = 1} (Ff)_1$$

$$= \sup_{\|f\|_\infty = 1} \gamma \sum_{i=1}^{n} (f, v_i) v_{i1} = \sup_{\|f\|_\infty = 1} \gamma \sum_{i=1}^{n} (f, v_i)$$

$$= \sup_{\|f\|_\infty = 1} \gamma \sum_{i=1}^{n} \sum_{k=1}^{2^n} f_k v_{ik} = \sup_{\|f\|_\infty = 1} \gamma \sum_{k=1}^{2^n} f_k \sum_{i=1}^{n} v_{ik}$$

$$= \gamma \sum_{k=1}^{2^n} \left\| \sum_{i=1}^{n} v_{ik} \right\| .$$

But now there are $2\binom{n}{0}$ ways that $|\Sigma_{i=1}^n v_{ik}| = n$, $2\binom{n}{1}$ ways that $|\Sigma_{i=1}^n v_{ik}| = n - 2$, $2\binom{n}{2}$ ways that $|\Sigma_{i=1}^n v_{ik}| = n - 4$, etc. Thus since $\gamma = \|v_1\|_2^2 = 2^{-n}$

$$\lambda(\ell_n^1) = 2[\binom{n}{0}n + \binom{n}{1}(n-2) + \binom{n}{2}(n-4) + \ldots + \binom{n}{r}s]/2^n$$

where $r = \frac{n-1}{2}$ and $s = 1$ if n is odd, and $r = \frac{n}{2} - 1$ and $s = 2$ if n is even. Using simple identities, one sees that this answer coincides with that of [9].

Also from Corollary 1 we obtain the following applications.

Recall that a basis $(e_i)_{i=1}^n$ is unconditional if $\|\sum_{i=1}^n a_i e_i\| = \|\sum_{i=1}^n |a_i| e_i\|$ for all choices of scalars $(a_i)_{i=1}^n$. Recall also that a basis is symmetric if $\|\Sigma\, a_i e_i\| = \|\Sigma |a_i| e_{\sigma(i)}\|$ for any sequence of scalars (a_i) and any permutation σ of $\{1,\dots,n\}$. (Thus of course a symmetric basis is in particular unconditional.)

THEOREM 4. Let E_n be an unconditional space and suppose $|v|$ has no local maxima in a determining "octant". Then for some constant r and some measure $\text{ext}|v|$, we have

$$P_{min}(E_n) = [r(\text{ext}|v|)v + (1 - r)\frac{v'_\perp}{w}].$$

Example 3 ([5]). $E_n = [1,t] \subset L^1[-1,1]$. Then it can easily be shown that $v(t) = (t, 1 - t^2)$ in the first (and determining) quadrant. Thus $v'_\perp/w = (2t,1)/w(t)$, where $w(t) = [1 + (t/\rho)^2]^{3/2}$, and $(\text{ext}|v|)v = (\delta_1 - \delta_{-1}, B\delta_0)$ where δ_z is the usual δ-function supported at z. Then $P_{min}(E_n) = [(u_1,u_2)] = [c_1(\delta_1 - \delta_{-1} + 2At/w(t)), c_2(B\delta_0 + A/w(t))]$ where c_1, c_2 are chosen so that $(v_i,u_j) = \delta_{ij}$, and minimizing over the parameters, we have $A = 1.562823\ldots$, $B = 2.297334\ldots$, $c_1 = .361381\ldots$, $c_2 = .265613\ldots$, $\rho = c_2/c_1$, and $\lambda(E_n) = \|P_{min}(E_n)\| = 1.22040491\ldots = \|Q\|$ where $Q : L^1[-1,1] \to E_n$ is a minimal projection ([8]). Thus the norm of the minimal projection from $L^1[-1,1]$ onto E_n as found in [8] is the absolute projection constant of E_n.

THEOREM 5. Let E_n be a symmetric space and suppose $|v|$ has no local maxima in a determining "sector". Then

$$P_{min}(E_n) = [r(\text{ext}|v|)v + (1 - r)\frac{v'_\perp}{w}]$$

where $(\text{ext}|v|)v$ is completely determined.

Example 4. $P_{min}(\ell_n^p) = [r(\Sigma\, \delta_{z_i})v + (1 - r)v'_\perp/w]$ where $\{z_i\}$ denote all corners of the ℓ_n^1-ball if $2 \le p \le \infty$ and of the ℓ_n^∞-ball if $1 \le p < 2$. Also $r = 1$ if $p = 1,\infty$ and $r = 0$ if $p = 2$. Furthermore it can be seen that $([v'_\perp/w], \|\cdot\|_\infty) \cong \ell_n^q$. In particular $\ell_2^p \cong ([v], \|\cdot\|_\infty)$ where

$$v = \frac{\cos t, \sin t}{(|\cos t|^q + |\sin t|^q)^{1/q}}, \qquad 0 \le t \le 2\pi$$

whence by symmetry

$$[v'_\perp/w] = [\frac{|\cos t|^{q-1}\mathrm{sgn}\,\cos t, |\sin t|^{q-1}\mathrm{sgn}\,\sin t}{(|\cos t|^q + |\sin t|^q)^{1/p}}] = [u]$$

and note that $([u], \|\cdot\|_\infty) \cong \ell_2^q$.

It is easy to conclude the following.

THEOREM 6. If $E_n \subset C(T), T \subset R^k$, then the $(n-1)$-dimensional surface $\tilde{v}$ is obtained by taking the surface of the convex hull of $\{e(t)\}_{t\in T} \cup -\{e(t)\}_{t\in T}$.

Theorem 6 in conjunction with Corollary 1 allows us to obtain the following examples.

Example 5. Consider $E_n = [(1, \sin t, \cos t)]$, $0 \le t \le 2\pi$ with $\|\ \|_\infty$. Then $\tilde{v}$ is the surface of a cylinder on the curved part of which $v'_\perp/w = (0, \sin t, \cos t)$ and on the top flat part $v'_\perp/w = (1,0,0)$, where $w(\cdot) \equiv 1$. $P_{min}(E_n)$ is then any nondegenerate linear combination of these two, as is well known. In this case the "discrete" solutions (1) and the "continuous" solutions (2) coincide.

Example 6. Consider $E_n = [(1, t, t^2)]$, $-1 \le t \le 1$ with $\|\ \|_\infty$. Then $P_{min}(E_n) = [u]$ has (up to symmetry through the origin) three "continuous" parts: $v'_\perp/w = (\frac{1+2|t|-t^2, 4\sqrt{2}t, -\sqrt{2}(|t|-1)^2}{w(t)})$ on part I, but parts II and III make no further contribution to the form for $v'_\perp/w$ and so $[v'_\perp/w] = [(t^2 - 2|t|, t, 1)/w(t)]$. Further $[u]$ has one "discrete" component, i.e., a measure with support at $e(\pm 1)$ and $e(0)$. From this $P_{min}(E_n)$ can be computed [6].

REFERENCES

[1] Chalmers, B. L., "The (*)-equation and the form of the minimal pro-

jection operator," in Approximation Theory IV (C.K. Chui, L.L. Schumaker, and J.D. Ward, eds.), pp.393-399.

[2] ______________, "A variational equation for minimal norm extensions," submitted for publication.

[3] ______________, "A natural simple projection with norm $\le \sqrt{n}$," J. Approx. Theory 32(1981) pp. 226-232.

[4] ______________, "The Fourier projection is minimal for regular polyhedral spaces," J. Approx. Theory, to appear.

[5] ______________, "The absolute projection constant for lines in $L^1[a,b]$," in preparation.

[6] Chalmers, B. L. and F. T. Metcalf, "The minimal projection onto the quadratics," in preparation.

[7] Chalmers, B. L. and B. Shekhtman, "Minimal projections and absolute projection constants for regular polyhedral spaces," Proc. Amer. Math. Soc., to appear.

[8] Franchetti, C. and E. W. Cheney, "Minimal projections in L^1-spaces," Duke Math. J. 43(1976), 501-510.

[9] Grünbaum, B., "Projection constants," Trans. Amer. Math. Soc. 95 (1960), 451-465.

MATRIX NORMS RELATED TO GROTHENDIECK'S INEQUALITY

A. M. Davie
Department of Mathematics
University of Edinburgh
James Clerk Maxwell Building
Mayfield Road
Edinburgh EH9 3JZ

Given positive integers n, p we define a norm $\|A\|_p$ for complex $n \times n$ matrices $A = (a_{ij})$ by

$$\|A\|_p = \sup\left\{\left|\sum_{i,j=1}^{n} a_{ij}\langle x_j, y_i\rangle\right| : x_j, y_i \in \mathbb{C}^p,\ \|x_j\| \le 1,\ \|y_i\| \le 1\right\} \quad (1)$$

where $\|x\|$ is the Euclidean norm on $\mathbb{C}^p$. We ask how this norm depends on p. Replacing $y_1, \ldots, y_n$ by their projections onto the span of $x_1, \ldots, x_n$ does not alter the inner products in (1), so we may assume that $x_1, \ldots, x_n, y_1, \ldots, y_n$ belong to a space of dimension n, which can be identified with $\mathbb{C}^n$. So $\|A\|_p = \|A\|_n$ for $p > n$. Clearly $\|A\|_r \le \|A\|_p$ for $r < p$.

Grothendieck's inequality asserts that $\|A\|_n \le K\|A\|_1$ where K is a constant independent of n, but it is not true that $\|A\|_n = \|A\|_1$ in general because Grothendieck's constant is greater than 1 (see e.g. [2, p.59]). One can ask: what is the smallest p such that $\|A\|_p = \|A\|_n$ for all $n \times n$ matrices A. We show here that $\|A\|_p = \|A\|_n$ if p is the integer part of $\sqrt{2n-1}$.

We first prove a simple Lemma.

<u>LEMMA</u> Let $1 \le m \le n$ and let M denote the set of Hermitian matrices of rank $n - m$. Then M is a manifold of real dimension $n^2 - m^2$.

<u>Proof</u> Let $H \in M$. We can write

$$H = U^{-1}\left[\begin{array}{c|c} 0 & 0 \\ \hline 0 & D \end{array}\right]U$$

where U is unitary and D is an invertible diagonal $(n-m) \times (n-m)$ matrix. Let V be the set of pairs (B, K) where B is a complex $m \times (n-m)$ matrix and K is a invertible $(n-m) \times (n-m)$ Hermitian matrix. Then V is an open subset of a real vector space of dimension $2m(n-m) + (n-m)^2 = n^2 - m^2$. Define $\phi : V \to n \times n$ matrices by

$$\phi(B,K) = U^{-1}\begin{bmatrix} BK^{-1}B^* & B \\ B^* & K \end{bmatrix} U \ .$$

Then $\phi(V)$ is a manifold of dimension $n^2 - m^2$, $\phi(V) \subseteq M$, and $\phi(V)$ contains a neighbourhood of H in M. So M is an $n^2 - m^2$ dimensional manifold.

<u>THEOREM</u> Let p be the integer part of $\sqrt{2n-1}$. Then $\|A\|_n = \|A\|_p$ for all $n \times n$ complex matrices A.

<u>Proof</u> Since the norms are continuous it suffices to prove the result for a dense set of A. We shall construct a nowhere dense set E of $n \times n$ matrices and prove that $\|A\|_n = \|A\|_p$ for $A \notin E$.

We note first that the expression to be maximized in (1) can be written as

$$\sum_{i=1}^{n} < \sum_{j=1}^{n} a_{ij}x_j ,\ y_i > \ ,$$

which shows that when the maximum is attained we have

$$\sum_j a_{ij}x_j = \lambda_i y_i \quad \text{and similarly} \quad \sum_i \bar{a}_{ij}y_i = \mu_j x_j \tag{2}$$

where $\lambda_i,\ \mu_i \geq 0$.

Moreover

$$\|A\|_n = \sup\{\sum_i \|\sum_j a_{ij}x_j\| : x_j \in \mathbb{C}^n,\ \|x_j\| \leq 1\}$$

$$= \sup\{\sum_j \|\sum_i \bar{a}_{ij}y_i\| : y_i \in \mathbb{C}^n,\ \|y_i\| \leq 1\}$$

Now define, for $k = 1,\ldots,n$,

$$F_k = \{A : \|A\|_n = \sup\{\sum_{i\neq k} \|\sum_j a_{ij}x_j\| : x_j \in \mathbb{C}^n,\ \|x_j\| \leq 1\}\}$$

$$G_k = \{A : \|A\|_n = \sup\{ \sum_{j\neq k} \| \sum_i \bar{a}_{ij} y_i \| : y_i \in \mathbb{C}^n, \|y_i\| \leq 1\}\}.$$

Then F_k and G_k are closed; we show that they have no interior. Suppose on the contrary that A is in the interior of F_k. Choose $x_1,\ldots x_n$ to maximize $\sum_{i\neq 1} \| \sum_j a_{ij} x_j \|$; we may suppose the x_j not all zero. Then since A is an interior point of F_k we can find $B \in F_k$ with $b_{ij}=a_{ij}$ for $i \neq k$, $j = 1\ldots n$ and $\sum b_{kj} x_j \neq 0$. But since $x_1,\ldots,x_n$ also maximizes $\sum_{i\neq k} \| \sum b_{ij} x_j \|$ this contradicts the definition of F_k. Similarly G_k has no interior.

Let $m = p+1$, so that $m > \sqrt{2n-1}$. Let J be the set of $n \times n$ matrices of the form $A = D_1 U(I-Y) D_2$ where

- D_1 is positive, diagonal with trace 1 (the set of such has real dimension $n-1$)
- U is unitary (the set of such has real dimension n^2)
- Y is Hermitian with rank $\leq n-m$ (the set of such has real dimension $n^2 - m^2$)
- D_2 is positive, diagonal (the set of such has real dimension n).

Note that the definition is unaltered if D_1 is not required to have trace 1 (D_1 can be made to have trace 1 by multiplying D_2 by a constant). The real dimension of J is not greater than the sum of the above dimensions, namely $2n^2 - m^2 + 2n - 1 < 2n^2$ since $m > \sqrt{2n-1}$. So J is nowhere dense. Hence $E = J \cup \cup_k (F_k \cup G_k)$ is nowhere dense.

To complete the proof we show that if $A \notin E$, and x_j, y_i satisfy (2), $x_1,\ldots,x_n$ span a space of (complex) dimension $\leq p$. If we let X be the $n \times n$ matrix whose columns are $x_1,\ldots,x_n$, similarly for Y, and denote by D_λ, D_μ the diagonal matrices with entries λ_i, μ_i then (2) can be written

$$AX = D_\lambda Y, \quad A^*Y = D_\mu X .$$

Since A does not belong to any F_k or G_k, $\lambda_1,\dots,\lambda_n$, $\mu_1,\dots,\mu_n$ are all strictly positive. Thus we can write

$$X = D_\mu^{-1}A^*Y = D_\mu^{-1}A^*D_\lambda^{-1}AX \quad \text{so} \quad (I-D_\mu^{-1}A^*D_\lambda^{-1}A)X = 0 .$$

Writing $B = D_\lambda^{-\frac{1}{2}}AD_\mu^{-\frac{1}{2}}$ we get $D_\mu^{-\frac{1}{2}}(I-B^*B)D_\mu^{\frac{1}{2}}X = 0$. Let $B = UP$ with U unitary, P nonnegative definite. Then $D_\mu^{-\frac{1}{2}}(I+P)(I-P)D_\mu^{\frac{1}{2}}X = 0$ and since $I + P$ is invertible, $(I-P)D_\mu^{\frac{1}{2}}X = 0$. Let $H = I - P$. Then H is Hermitian and $A = D_\lambda^{\frac{1}{2}}BD_\mu^{\frac{1}{2}} = D_\lambda^{\frac{1}{2}}U(I-H)D_\mu^{\frac{1}{2}}$ and since $A \notin J$ it follows that $\text{rank}(Y) > n - m$. But $HD_\mu^{\frac{1}{2}}X = 0$ and $D_\mu^{\frac{1}{2}}$ is invertible, so $\text{rank}(X) < m$, and finally $\text{rank}(X) \le p$ which is the desired result.

<u>COROLLARY</u> Let A be a complex 2×2 matrix. Then $\|A\|_1 = \|A\|_2$. This has been proved by different means by Haagerup [3].

<u>Remark 1</u> A similar analysis can be carried out in the real case - i.e. A is restricted to have real entries and the maxima are taken over x_j, $y_j \in \mathbb{R}^n$. The orthogonal matrices form a set of dimension $\frac{1}{2}n(n-1$, and the symmetric matrices of rank $n - m$ form a set of dimension $\frac{1}{2}(n(n+1) - m(m+1))$, so the requirement on m is $m(m+1) > 4n - 2$. This gives $\|A\|_n = \|A\|_2$ for 2×3 matrices.

<u>Remark 2</u> Our results can be interpreted in terms of the maximal rank of extreme points of certain convex sets of matrices. Let W_n denote the set of matrices $B = (b_{ij})$ of the form $b_{ij} = \langle x_i, y_j\rangle$ where $x_1,\dots,x_n$, $y_1,\dots,y_n \in \mathbb{C}^n$, $\|x_i\| \le 1$. Let Z_n denote the set of matrices of the form $b_{ij} = \langle x_i, x_j\rangle$, $x_1,\dots,x_n \in \mathbb{C}^n$, $\|x_i\| = 1$. Then Z_n is the set of $n \times n$ nonnegative Hermitian matrices whose diagonal elements are all 1, and so Z_n is a compact convex set. W_n is the set of matrices B such that there exists a matrix of the form

$$\begin{bmatrix} H & B \\ B^* & K \end{bmatrix}$$

in Z_{2n}, so W_n is also compact and convex. It is not hard to see that the smallest K for which $\|A\|_n = \|A\|_k$ for all $n \times n$ matrices A coincides with the maximum rank of the extreme points of W_n, so our theorem states that the rank of any extreme point of W_n is $\le \sqrt{2n-1}$. It is known [1] that the extreme points of Z_n have rank $\le \sqrt{n}$ from which one can conclude that the extreme points of W_n have rank $\le \sqrt{2n}$ but I do not know if one can get $\sqrt{2n-1}$ by this type of argument. Also one can show that Z_n has extreme points of rank $[\sqrt{n}]$ (where [ ] denotes integer part) but I do not know whether W_n has extreme points of rank $[\sqrt{2n-1}]$.

REFERENCES

[1] J.P.R. Christensen and J. Vesterstrøm, A note on extreme positive definite matrices, Math. Ann. 244(1979) 65-68.

[2] A. Grothendieck, Résumé de la théoriè métrique des produits tensoriels topologiques, Bol. Soc. Mat. Brasil, Sao Paulo 8(1956) 1-79.

[3] U. Haagerup, personal communication.

CHARACTERIZATION OF WEAK COMPACTNESS IN FUNCTION SPACES BY MEANS OF UNIFORM CONVERGENCE OF EXTENDED OPERATORS

Nicolae Dinculeanu
Department of Mathematics
University of Florida
Gainesville, Florida 32611

1. Introduction

In [9] we gave a new characterization of weak compactness in $L^p(\mu)$, in terms of uniform weak convergence of "admissible" sequences of operators (see definition 2). More generally, admissible sequences of operators can be used to characterize conditional σ'-compactness in $L_E^p(\mu)$, where E is an arbitrary Banach space and $\sigma' = (L_E^p , L_{E'}^q)$; or to characterize conditional weak compactness in $L_E^p(\mu)$ in case E' has the RNP (Radon Nikodym property). A set $A \subset L_E^p$ is said to be conditionally compact for a certain topology τ if any sequence from A contains a Cauchy subsequence.

An important example of an admissible sequence of operators is that of the conditional expectations determined by a sequence of finite partitions generating the whole σ-algebra Σ, in case Σ is separable. But if Σ is not separable, the conditional expectations determined by the finite partitions form an admissible net, which has no cofinal sequence.

The following problem arises: can weak compactness in $L^p(\mu)$ be characterized by uniform weak convergence of admissible nets (rather that sequences)? More generally, can admissible nets be used to characterize conditional σ'-compactness in L_E^p ?

In [9] we proved that the answer is positive in case E' has the RNP. We remark that in this case, σ' is the weak topology. In particular, for E = R, this gives a characterization of relative weak compactness in $L^1(\mu)$ by means of admissible nets.

In this paper we shall prove that the answer to the above problem is still positive in case E is an arbitrary Banach space but the net consists of "extended" operators (see §3 below). In particular, this is the case of nets of conditional expectations, or of convolutions with the elements of an approximate unit in $L_E^p(G)$ or of translations in $L^p_E(G)$, where G is an abelian locally compact group endowed with a Haar measure.

At the same time we prove that admissible nets of extended operators can be used to characterize uniformly σ-additive sets in L^1_E.

We apply then the above-mentioned result to give a characterization of weak compactness in terms of uniform convergence of conditional expectations.

2. The Main Result

Let (X,Σ,μ) and (X',Σ',ν) be two strictly localizable measure spaces [12], E and F two Banach spaces. Let Σ_f be the δ-ring of sets $A \in \Sigma$ with $\mu(A) < \infty$; similar definition for Σ'_f. If $1 \leq p < \infty$ and $T: L^p_E(\mu) \to L^p_f(\nu)$ is a linear operator we set $\|T\|_p = \sup\{\|Tf\|_p;\ f \in L^p_E(\mu),\ \|f\|_p \leq 1\}$ and $\|T\|_\infty = \sup\{\|Tf\|_\infty;\ f \in L^1_E(\mu) \cap L^\infty_E(\mu),\ \|f\|_\infty \leq 1\}$.

Definition 1. We say that an operator $T: L^p_E(\mu) \to L^p_F(\nu)$ has the finite measure property (FMP) on a class of sets $R \subset \Sigma_f$, if, for every set $C \in R$ there is a set $\phi(C,T) \in \Sigma'_f$ such that, if $f \in L^p_E(\mu)$ vanishes μ-a.e. outside C, then Tf vanishes ν-a.e. outside $\phi(C,T)$.

Evidently, if $\nu(X') < \infty$, any operator T has the FMP on Σ_f.

Definition 2. Let I be a directed set, $(T_\alpha)_{\alpha \in I}$ a net of continuous linear operators from $L^p_E(\mu)$ into $L^p_F(\nu)$ and $T_\beta: L^p_E(\mu) \to L^p_F(\nu)$ a continuous linear operator. We say that the net (T_α) is admissible and has limit T_β if the following conditions are satisfied:

In case $p = 1$:

1.) $\sup_\alpha \|T_\alpha\|_1 < \infty$ and $\sup_\alpha \|T_\alpha\|_\infty < \infty$;
2.) for each $\alpha \in I$, the adjoint T^*_α maps $L^1_{F'}(\nu) \cap L^\infty_{F'}(\nu)$ into $L^1_{E'}(\mu) \cap L^\infty_{E'}(\mu)$ and for every $g \in L^1_{F'}(\nu) \cap L^\infty_{F'}(\nu)$ we have $\lim_\alpha T^*_\alpha g = T^*_\beta g$, strongly in $L^1_{E'}(\mu)$;
3.) There is a class $R \subset \Sigma_f$ generating Σ_f such that each T_α has the FMP on R, and such that if, for every $C \in R$ we denote $C_\alpha = \phi(C,T_\alpha)$, then $\lim_\alpha \phi_{C_\alpha} = \phi_{C_\beta}$

strongly in $L^1(\nu)$. (In case $\nu(X') < \infty$ condition 3 is superfluous.)

In case $1 < p < \infty$:

1') $\sup_\alpha \|T_\alpha\|_p < \infty$;

2') for every $\alpha \in I$, we have $T_\alpha^* L_{F'}^q(\nu) \subset L_{E'}^q(\mu)$ and for every $g \in L_{F'}^q(\nu)$ we have $\lim_\alpha T_\alpha^* g = T_\beta^* g$, strongly in $L_{E'}^q(\mu)$.

Examples. 1.) Let $(\Sigma_\alpha)_{\alpha \in I}$ be a directed family of sub δ-rings of Σ_f and Σ_β the δ-ring generated by all Σ_α. Let E_α and E_β be the conditional expectation operators corresponding to Σ_α and Σ_β respectively. Then the net of operators (E_α) from $L_E^p(\mu)$ into $L_E^p(\mu)$ is admissible and has limit E_β for $1 < p < \infty$. For $p = 1$, (E_α) is admissible and has limit E_β iff condition 3 in definition 2 is satisfied. In particular, condition 3 is satisifed if $\mu(X) < \infty$ and each Σ_α is generated by a finite partition π_α or if $\nu(X') < \infty$.

2.) Let G be a locally compact abelian group, μ a Haar measure on G and (u_V) an approximate unit, where V runs over a base of relatively compact neighborhoods of 0 in G, ordered downwards by inclusion. For each V, let T_V be the convolution operator defined for $f \in L_E^p(\mu)$ by $T_V f = f * u_V$. Then the net (T_V) is admissible on $L_E^p(\mu)$ and has limit the identity operator I (see [7]).

3.) Let G and μ be as in example 2 and let (h_α) be a net of elements of G converging to 0 in G. For each α let T^α be the translation operator defined for $f \in L_E^p(\mu)$ by $(T^\alpha f)(x) = f(x+h_\alpha)$ for $x \in G$. Then the net (T^α) is admissible on $L_E^p(\mu)$ and has limit the identity operator I (see [7]).

Definition 3. A linear operator $S: L_E^p(\mu) \to L_E^p(\nu)$ is called an extended operator if there exists a linear operator $T: L^p(\mu) \to L^p(\nu)$ such that $\langle Sf, x'\rangle = T(\langle f, x'\rangle)$ ν-a.e. for $f \in L_E^p(\mu)$ and $x' \in E'$. S is called the extension of T and is denoted by T_E.

The conditional expectations, the convolution operators and the translation operators in the preceding examples are all extended operators.

We can now state our main result.

Theorem 4. Let $(T_{\alpha E})_{\alpha \in I}$ be an admissible net of extended continuous linear operators from $L_E^p(\mu)$ into itself, $1 < p < \infty$, with limit the identity operator I on $L_E^p(\mu)$.

A.) A set $K \subset L^p_E(\mu)$ is conditionally σ'-compact iff for every separable subset $K_0 \subset K$ there exists an increasing sequence (α_n) such that all sets $T_{\alpha_n E} K_0$ are conditionally σ'-compact and $\lim_n T_{\alpha_n E} f = f$ in $L^p_E(\mu)$ for the σ'-topology, uniformly for $f \in K_0$.

B.) Let $K \subset L^1_E(\mu)$ be bounded. The set $|K| = \{|f|;\ f \in K\}$ is uniformly σ-additive iff for any separable subset $K_0 \subset K$, there is an increasing sequence (α_n) such that each $|T_{\alpha_n E} K_0|$ is uniformly σ-additive and $\lim_n T_{\alpha_n E} f = f$ in $L^1_E(\mu)$ for the σ'-topology, uniformly for $f \in K_0$.

If E' has the RNP then σ' should be replaced with the weak topo logy. If, moreover, $E = R$, then "conditionally σ'-compact" should be replaced with "relatively weakly compact". The proof of the theorem is done in section 4.

We shall use the following result proved in [9] for admissible sequences (rather than nets) of (not necessarily extended) operators.

Theorem 5. Let (T_n) be an admissible sequence of operators from $L^p_E(\mu)$ into itself, with limit I, the identity operator of $L^p_E(\mu)$.

A.) A set $K \subset L^p_E(\mu)$ is conditionally σ'-compact iff each $T_n K$ is conditionally σ'-compact and $\lim_n T_n f = f$ in $L^p_E(\mu)$, for the σ'-topology, uniformly for $f \in K$.

B.) Let $K \subset L^1_E(\mu)$ be bounded. The set $|K| = \{|f|;\ f \in K\}$ is uniformly σ-additive iff each $|T_n K|$ is uniformly σ-additive and $\lim_n T_n f = f$ in $L^1_E(\mu)$ for the σ'-topology, uniformly for $f \in K$.

3. Extended Operators

In this section we state the properties of extended operators which will be used in the proof of theorem 4.

Let $T_E: L^p_E(\mu) \to L^p_E(\nu)$, $1 \leq p < \infty$, be a continuous extended operator, i.e.

1.) $\langle T_E f\ ,\ x' \rangle = T \langle f\ ,\ x' \rangle$, ν-a.e., for $f \in L^p_E(\mu)$, $x' \in E'$;

The following properties are easy to prove:

2.) $T_E(x\phi) = x\ T(\phi)$, ν-a.e., for $\phi \in L^p(\mu)$, $x \in E$;

3.) T_E has the FMP on a class $R \subset \Sigma_f$ iff T has the FMP on R, and we can take $\phi(C,T_E) = \phi(C,T)$ for $C \in R$;

4.) $\|T\|_p \le \|T_E\|_p$;

5.) Let $\Sigma_0 \subset \Sigma$ and $\Sigma_0' \subset \Sigma'$ be sub σ-algebras. Then
$T_E\, L_E^p(\Sigma_0,\mu) \subset L_E^p(\Sigma_0',\nu)$ iff $T\, L^p(\Sigma_0,\mu) \subset L^p(\Sigma_0',\nu)$.

Let q be the conjugate number of p. The dual of $L_E^p(\mu)$ is isomorphic to the space $L_{E'}^q[E,\mu]$ of (equivalence classes of) functions g: $X \to E'$ such that $\langle x,g\rangle$ is Σ-measurable for every $x \in E$ and $\|g\|_q < \infty$ (see [12]). We have $L_{E'}^q(\mu) \subset L_{E'}^q[E,\mu]$, with equality if E' has the RNP.

Consider the transposed operator $(T_E)^*: L_{E'}^q[E,\nu] \to L_{E'}^q[E,\mu]$. We shall assume that $(T_E)^*\, L_{E'}^q(\nu) \subset L_{E'}^q(\mu)$. Then we have the following properties:

6.) $\langle x\,,\,(T_E)^*g\rangle = T^*(\langle x,g\rangle)$, μ-a.e. for $g \in L_{E'}^q(\nu)$ and $x \in E$;

7.) $(T_E)^*(x'\phi) = x'T^*(\phi)$, μ-a.e. for $\phi \in L^q(\nu)$ and $x' \in E'$.

8.) T_E is continuous for the topologies $\sigma(L_E^p(\mu),\, L_{E'}^q(\mu))$ and $\sigma(L_E^p(\nu),\, L_{E'}^q(\nu))$ respectively on $L_E^p(\mu)$ and $L_E^p(\nu)$.

9.) Let $\Sigma_0 \subset \Sigma$ and $\Sigma_0' \subset \Sigma'$ be sub σ-algebras. Then
$(T_E)^*\, L_{E'}^q(\Sigma_0',\nu) \subset L_{E'}^q(\Sigma_0,\mu)$ iff $T^*L^q(\Sigma_0',\nu) \subset L^q(\Sigma_0,\mu)$.

For p = 1 we have additional properties:

10.) Any continuous linear operator $T: L^1(\mu) \to L^1(\nu)$ has an extension $T_E: L_E^1(\mu) \to L_E^1(\nu)$ and we have $\|T_E\|_1 = \|T\|_1$ (see [10]).

11.) We have $T(L^1(\mu) \cap L^\infty(\nu)) \subset L^1(\mu) \cap L^\infty(\nu)$ and $\|T\|_\infty < \infty$ iff $T_E(L_E^1(\mu) \cap L_E^\infty(\nu)) \subset L_E^1(\mu) \cap L_E^\infty(\nu)$ and $\|T_E\|_\infty < \infty$. In this case $\|T_E\|_\infty = \|T\|_\infty$.

12.) Assume $T(L^1(\mu) \cap L^\infty(\mu)) \subset L^1(\nu) \cap L^\infty(\nu)$ and $\|T\|_\infty < \infty$.

Then $(T_E)^*\, (L_{E'}^1(\nu) \cap L_{E'}^\infty(\nu)) \subset L_{E'}^1(\mu) \cap L_{E'}^\infty(\mu)$ and $\|(T_E)^*\|_1 = \|T_{E'}\|_\infty$.

13.) If $T(L^1(\mu) \cap L^\infty(\mu)) \subset L^1(\nu) \cap L^\infty(\nu)$ and $\|T\|_\infty < \infty$, then T^* can be extended from $L_{E'}^1(\nu)$ into $L_{E'}^1(\mu)$ and the extension $(T^*)_{E'}$ satisfies:
$(T^*)_{E'}(L_{E'}^1(\nu) \cap L_{E'}^\infty(\nu)) \subset L_{E'}^1(\mu) \cap L_{E'}^\infty(\mu)$ and
$(T^*)_{E'} = (T_E)^*$ on $L_{E'}^1(\nu) \cap L_{E'}^\infty(\nu)$.

<u>Proof of property 12</u>. Let $f \in L^1(\nu) \cap L^\infty(\nu)$. Then $T^* f \in L^\infty(\mu)$.

a.) We prove first that $b = \sup\{\int |(T^*f)g| d\mu;\ g \in L^1(\mu) \cap L^\infty(\mu),\ \|g\|_\infty \leq 1\} < \infty$. In fact, let $g \in L^1(\mu) \cap L^\infty(\mu)$ with $\|g\|_\infty \leq 1$ and set $h = \operatorname{sgn}(T^*f)g$. Then $gh \in L^1(\mu) \cap L^\infty(\mu)$, hence $\int |(T^*f)g| d\mu = \int (T^*f)gh d\mu = \int fT(gh)d\nu \leq$ $\leq \|f\|_1 \|T\|_\infty$ therefore $b \leq \|f\|_1 \|T\|_\infty < \infty$.

b.) We prove next that there is a set $A \in \Sigma$ of σ-finite measure such that T^*f vanishes μ-a.e. outside A and $T^*f \in L^1(\mu) \cap L^\infty(\mu)$. In fact, set $c = \sup\{\int_A |T^*f| d\mu;\ \mu(A) < \infty\}$ and let (A_n) be an increasing sequence of Σ_f such that $c = \sup_n \int_{A_n} |T^*f| d\mu$. Taking $g = \phi_{A_n}$ in step a) we deduce that $\sup_n \int_{A_n} |T^*f| d\mu \leq b < \infty$. The set $A = \cup A_n$ has σ-finite measure and $\phi_A |T^*f|$ is μ-integrable. Let $A_0 \in \Sigma_f$ be a set disjoint from A such that $T^*f \neq 0$ on A_0. Then $c + \int_{A_0} |T^*f|\, d\mu =$ $\sup_n \int_{A_n} |T^*f| d\mu + \int_{A_0} |T^*f| d\mu = \sup_n \int_{A_n \cup A_0} |T^*f|\, d\mu \leq c$. It follows that $\int_{A_0} |T^*f|\, d\mu = 0$, hence $\mu(A_0) = 0$. We deduce then that $|T^*f| = \phi_A |T^*f|$, μ-a.e. therefore $T^*f \in L^1(\mu)$. Then a simple computation shows that $\|T^*\|_1 = \|T\|_\infty$.

4. Admissible Nets of Extended Operators

Proposition 6. If $(T_{\alpha E})_{\alpha \in I}$ is an admissible net of extended operators from $L^p_E(\mu)$ into $L^p_E(\nu)$, $1 \leq p < \infty$, with limit an extended operator $T_{\beta E}$, then $(T_\alpha)_{\alpha \in I}$ is an admissible net of operators from $L^p(\mu)$ into $L^p(\nu)$ with limit T_β. If $p = 1$, the converse implication is also true.

Proof. Consider first the case $1 < p < \infty$. Condition 1') of definition 2 for (T_α) follows from the hypothesis and property 4. To prove condition 2') of definition 2 we use property 9 (with $\Sigma_0 = \Sigma$ and $\Sigma_0' = \Sigma'$) to deduce that $T^*_\alpha L^q(\Sigma', \nu) \subset L^q(\Sigma, \mu)$. Then for $g \in L^q(\Sigma', \nu)$ and $x' \in E'$ with $|x'| = 1$ we have, by property 7,
$\|T^*_\alpha g - T^*_\beta g\|_q = \|x' T^*_\alpha g - x' T^*_\beta g\|_q = \|(T_{\alpha E})^*(x'g) - (T_{\beta E})^*(x'g)\|_q$
hence, $\lim_\alpha T^*_\alpha g = T^*_\beta g$ in $L^q(\mu)$.

Consider now the case $p = 1$. By properties 11 and 3 conditions 1 and 3 of definition 2 are satisfied for (T_α) if and only if they are satisfied for $(T_{\alpha E})$. By properties 11 and 12 we have

$(T_{\alpha E})^* (L^1_{E'}(\nu) \cap L^\infty_{E'}(\nu)) \subset L^1_{E'}(\mu) \cap L^\infty_{E'}(\mu)$ iff $T^*_\alpha (L^1(\nu) \cap L^\infty(\nu)) \subset L^1(\mu) \cap L^\infty(\mu)$. If $(T_{\alpha E})$ satisfies condition 2 of definition 2, then by the above computations, with q replaced by 1, we deduce that (T_α) also satisfies condition 2. Conversely, assume that $\lim_\alpha T^*_\alpha \phi = T^*_\beta \phi$ in $L^1(\mu)$ for $\phi \in L^1(\nu) \cap L^\infty(\nu)$. Let $g \in L^1_{E'}(\nu) \cap L_{E'}(\nu)$. If $g = x'\phi$ with $x' \in E'$ and $\phi \in L^1(\nu) \cap L^\infty(\nu)$ then using property 7 we deduce that $\lim_\alpha (T_{\alpha E})^* g = (T_{\beta E})^* g$ in $L^1_{E'}(\mu)$. Then this equality is valid if g is a step function. Finally, the equality is valid for an arbitrary g, since $\sup_\alpha \|(T_{\alpha E})^*\|_1 = \sup_\alpha \|T^*_\alpha\|_1 < \infty$ and we can apply the Banach Steinhauss theorem.

<u>Proposition 7.</u> <u>Let $(T_\alpha)_{\alpha \in I}$ be an admissible net of continuous linear operators from $L^p(\mu)$ into $L^p(\nu)$ with limit T_β and let $\Sigma_0 \subset \Sigma$ be a separable sub σ-algebra.</u>

<u>Then there are separable σ-algebras $\Sigma_\infty \subset \Sigma$ and $\Sigma'_\infty \subset \Sigma'$, with $\Sigma_0 \subset \Sigma_\infty$ and an increasing sequence (α_n) of indices such that the restrictions of the T_{α_n} to $L^p(\mu)$ form an admissible sequence of operators from $L^p(\Sigma_\infty, \mu)$ into $L^p(\Sigma'_\infty, \nu)$, with limit T_β restricted to $L^p(\Sigma_\infty, \mu)$.</u>

<u>Proof</u>. a.) If $p = 1$, let R be a countable class of sets of finite measure generating Σ_0 and satisfying condition 3 of definition 2. By a diagonal process we can find an increasing sequence $(\alpha_{0,n})$ with $\alpha_{0,0} = \beta$ such that, for any sequence $\alpha_n > \alpha_{0,n}$ we have $\lim_n \phi_{C_{\alpha_n}} = \phi_{C_\beta}$ in $L^1(\Sigma', \nu)$ for each $C \in R$.

If $p > 1$, we start with any sequence $(\alpha_{0,n})$ with $\alpha_{0,0} = \beta$.

b.) The set $A_0 = \{T_{\alpha_{0,n}} L^p(\Sigma_0, \mu);\ n > 0\}$ is separable in $L^p(\Sigma', \nu)$. There is a countably generated σ-algebra $\Sigma'_0 \subset \Sigma'$ such that $A_0 \subset L^p(\Sigma'_0, \nu)$.

c.) Since $L^1(\Sigma'_0, \nu) \cap L^\infty(\Sigma'_0, \nu)$ is separable in $L^1(\nu)$ for $p = 1$, and $L^q(\Sigma'_0, \nu)$ is separable for $1 < p < \infty$, by a diagonal process we can find an increasing sequence $(\alpha_{1,n})$ with $\alpha_{1,n} > \alpha_{0,n}$ and $\alpha_{1,0} = \beta$ such that for any sequence $\alpha_n > \alpha_{1,n}$ we have $T^*_{\alpha_n} g \to T^*_\beta g$ as $n \to \infty$, in $L^1(\mu) \cap L^\infty(\mu)$ for $g \in L^1(\Sigma'_0, \nu) \cap L^\infty(\Sigma'_0, \nu)$ in case $p = 1$, respectively in $L^q(\mu)$ for $g \in L^q(\Sigma'_0, \nu)$ in case $1 < p < \infty$.

d.) The set $B_1 = \{T^*_{\alpha_{i,n}} L^1(\Sigma'_0, \nu) \cap L^\infty(\Sigma'_0, \nu);\ i = 1,2,\ n > 0\}$ is separable in $L^1(\mu)$ in case $p = 1$, and the set $B_p = \{T^*_{\alpha_{i,n}} L^q(\Sigma'_0, \nu);\ i = 1,2,\ n > 0\}$ is separable in $L^q(\nu)$ in case

$1 < p < \infty$. There is a countably generated σ-algebra Σ_1 with $\Sigma_0 \subset \Sigma_1 \subset \Sigma$ such that $B_1 \subset L^1(\Sigma_1,\mu)$, respectively $B_p \subset L^q(\Sigma_1,\mu)$.

e.) By induction we can find an increasing sequence (Σ_k) of separable sub σ-algebras of Σ containing Σ_0, an increasing sequence (Σ_k') of separable sub σ-algebras of Σ', and an increasing sequence $(\alpha_{i,n})_{n>0}$ for $i > 0$, such that: $\alpha_{i,0} = \beta$, $\alpha_{i,n} < \alpha_{i+1,n}$ and such that, in case $p = 1$, for $i < k$ and $n > 0$ we have

$$T_{\alpha_{i,n}} L^1(\Sigma_k,\mu) \cap L^\infty(\Sigma_k,\mu) \subset L(\Sigma_k',\nu) \cap L^\infty(\Sigma_k',\nu),$$

$$T^*_{\alpha_{i,n}} L^1(\Sigma_k',\nu) \cap L^\infty(\Sigma_k',\nu) \subset L^1(\Sigma_k,\mu) \cap L^\infty(\Sigma_k,\mu)$$

and $\lim_n T^*_{\alpha_{i,n}} g = T^*_\beta g$ in $L^1(\mu)$ for $g \in L^1(\Sigma_k',\nu) \cap L^\infty(\Sigma_k',\nu)$; respectively, in case $1 < p < \infty$, for $i < k$ and $n > 0$ we have $T_{\alpha_{i,n}} L^p(\Sigma_k,\mu) \subset L^p(\Sigma_k',\nu)$, $T^*_{\alpha_{i,n}} L^q(\Sigma_k', \nu) \subset L^q(\Sigma_k,\mu)$ and $\lim_n T^*_{\alpha_{i,n}} g = T^*_\beta g$ in $L^q(\Sigma_k,\mu)$ for $g \in L^q(\Sigma_k',\nu)$.

f.) We take Σ_∞ and Σ_∞' the σ-algebras generated by $\cup\Sigma_k$ and $\cup\Sigma_k'$ respectively, and $\alpha_n = \alpha_{n,n}$ and the conclusion of the proposition follows.

<u>Remark</u>. If $X = X'$ and $\Sigma = \Sigma'$ we can take $\Sigma_0 \subset \Sigma_0' \subset \Sigma_1 \subset \Sigma_1' \subset \ldots$ hence, $\Sigma_\infty = \Sigma_\infty'$.

<u>Proof of Theorem 4</u>. Let $K \subset L^p_E(\mu)$ be conditionally σ'-compact and let $K_0 \subset K$ be a separable subset. Let $\Sigma_0 \subset \Sigma$ be a separable sub σ-algebra such that all functions of K_0 are Σ_0 - measurable. By the preceding proposition, there is a separable σ-algebra Σ_∞ with $\Sigma_0 \subset \Sigma_\infty \subset \Sigma$, and an increasing sequence (α_n) such that (T_{α_n}) is an admissible sequence from $L^p(\Sigma_\infty,\mu)$ into itself with limit I. Then $K_0 \subset L^p(\Sigma_\infty,\mu)$ and $(T_{\alpha_n E})$ is an admissible sequence from $L^p_E(\Sigma_\infty,\mu)$ into itself. This follows from proposition 6 for $p = 1$. For $1 < p < \infty$, condition 1') of definition 2 is satisfied by $(T_{\alpha_n E})$ since it is satisifed by $(T_{\alpha E})$. We have $(T_{\alpha_n E})^* L^q_{E'}(\Sigma_\infty,\mu) \subset L^q_{E'}(\Sigma_\infty,\mu)$ by proposition 7 and property 9, and for $g \in L^q_{E'}(\Sigma_\infty,\mu)$ we have $\lim_\alpha (T_{\alpha_n E})^* g = g$ in $L^q_{E'}(\mu)$, first for step functions using property 7, then for all $g \in L^q_{E'}(\Sigma_\infty,\mu)$ by the Banach Steinhauss theorem. We can now apply theorem 5 and deduce that $\lim_n T_{\alpha_n E} f = f$ in $L^p_E(\Sigma_\infty,\mu)$ for the topology $\sigma(L^p_E(\Sigma_\infty,\mu), L^q_{E'}(\Sigma_\infty,\mu))$, uniformly for

$f \in K_0$, and we notice that the above topology is the restriction of the topology σ' to the subspace $L_E^p(\Sigma_\infty,\mu)$. The converse implication follows from theorem 5 since a set is conditionally σ'-compact iff any separable subset is conditionally σ'-compact.

5. Application to Nets of Conditional Expectations

We can apply the above result to a net of conditional expectations, determined by an increasing net of sub σ-algebras of Σ, for example, a filtration $(F_t)_{t>0}$. In the case of conditional expectations E_π determined by finite "partitions" π of X into sets from Σ_f we obtain the following improvements of results in [4] and [5].

Theorem 8. Assume (X,Σ_∞,μ) is separable and let (π_n) be an increasing sequence of finite partitions cofinal to the net (π) of all finite partitions over Σ_f.

A.) A set $K \subset L_E^p(\mu)$ is conditionally σ'-compact iff the following conditions are satisfied:

1.) for every set $A \in \Sigma_f$ the set $K(A) = \{\int_A f d\mu;\ f \in K\}$ is conditionally weakly compact in E;

2.) $\lim_n E_{\pi_n} f = f$, in $L_E^p(\mu)$, for the σ'-topology, uniformly for $f \in K$.

B.) Let $K \subset L_E^1(\mu)$ be bounded. The set $|K| = \{|f|;\ f \in K\}$ is uniformly σ-additive iff condition 2 above is satisifed.

Proof. Let $K \in L_E^p(\mu)$ and assume conditions 1) and 2) of the statements are satisfied. Let $\pi = (A_1,\dots,A_n)$ be a finite partition consisting of sets $A_i \in \Sigma$ with $0 < \mu(A_i) < \infty$ and let E_π be the corresponding conditional expectation: $E_\pi f = \Sigma_i[\mu(A_i)^{-1} \int_{A_i} fd\mu]\ \phi_{A_i}]$ for $f \in L_E^p(\mu)$. Then $E_\pi K$ is finite dimensional in $L_E^p(\mu)$, hence it is relatively compact. By theorem 5, K is conditionally σ'-compact. The converse implication of A) follows from theorem 5 and from the continuity of the mapping $f \to \int_A fd\mu$ of $L_E^p(\mu)$ into E.

To prove B, assume $|K|$ is bounded in $L^1(\mu)$. Then, for every $A \in \Sigma$ we have

$$\int_A |Ef|d\mu = \Sigma_i\ \mu(A_i)^{-1}\ |\int_A fd\mu|\ \mu(A \cap A_i) \leq M\Sigma_i\ \mu(A_i)^{-1}\ \mu(A \cap A_i),$$

where $M = \sup\{\|f\|_1;\ f \in K\}$. We deduce that each $|E_\pi K|$ is uniformly σ-additive; then B) follows from theorem 5.

A combination of theorems 4 and 8 yields:

Theorem 9. A.) *A set* $K \subset L^p_E(\mu)$ *is conditionally weakly compact iff for every separable subset* $K_0 \subset K$ *there is an increasing sequence* (π_n) *of finite partitions satisfying conditions* 1) *and* 2) *of theorem 8, for* K *replaced with* K_0.

B.) *Let* $K \in L^1_E(\mu)$ *be bounded. The set* $|K|$ *is uniformly* σ-*additive iff for every separable subset* $K_0 \subset K$, *there is an increasing sequence* (π_n) *of finite partitions satisfying condition 2 of theorem 8, for* $f \in K_0$.

Bibliography

1. J. Batt, On weak compactness in spaces of vector valued measures and Bochner integrable functions in connection with the Radon Nikodym property of Banach spaces, Revue Roumaine Math. Pures et Appl. 19 (1974), 285-304.

2. J. Batt and N. Dinculeanu, On the weak compactness criteria of Kolmogorov-Tamarkin and M. Riesz type in the space of Bochner integrable functions over a locally compact group, Measure Theory and Applications, Proceedings, Sherbrook-Canada; Springer Lecture Notes 1033 (1983), 43-58.

3. J. Bourgain, An averaging result for ℓ^1-sequences and applications to conditionally weakly compact sets in L^1_X, Israel J. Math. 32 (1979), 289-298.

4. J.K. Brooks and N. Dinculeanu, Weak Compactness in spaces of Bochner integrable functions and applications, Advances in Math 24 (1977), 172-188.

5. __________, Conditional expectations and weak and strong compactness in spaces of Bochner integrable functions, J. Multivariate Analysis, 9 (1979), 420-427.

6. N. Dinculeanu, Uniform σ-additivity and uniform convergence of conditional expectations in the space of Bochner or Pettis integrable functions, General Toplogy and Modern Analysis, Academic Press (1981), 391-397.

7. __________, On Kolmogorov-Tamarkin and M. Riesz compactness criteria in function spaces over a locally compact group, J. Math. Analysis and Appl. 89 (1982), 67-85.

8. __________, Uniform σ-additivity in spaces of Bochner or Pettis integrable functions over a locally compact group, Proc. Amer. Math. Soc. 87 (1983), 627-633.

9. __________, Weak compactness and uniform convergence of operators in spaces of Bochner integrable functions, J. Math. Analysis and Appl. (to appear).

10. N. Dinculeanu and C. Ionescu Tulcea, Extensions of certain operators to spaces of abstract integrable functions, Rendiconti del Circolo Mat. Palermo, 31 (1982), 433-448.

11. A. Grothendrick, Sur les applications lineáires faiblement compactes d'espaces du type C(K), Canad. J. Math. 5 (1983), 129-173.

12. A. and C. Ionescu Tulcea, Topics in the theory of lifting, Springer, (1969).

13. A. Kolmogorov, Ueber die Kompaktheit der Funktionenmengen bei der Konvergenz in Mittel, Nachr. Acad. Wiss. Göttingen Math.-Phys. Kl. II (1931), 60-63.

14. M. Nicolescu, Analiza Matematica, vol III. Ed. Technica, Bucarest, 1960.

15. M. Riesz, Sur les ensembles compacts de functions sommables, Acta Litt. Sci. Univ. Szeged 6 (1933), 136-142.

16. J.D. Tamarkin, On the compactness of the space L, Bull. Amer. Math. Soc. 38 (1932), 79-84.

17. A. Weil, L'Intégration dans les groupes topologigues et ses applications, Hermann, Paris, 1953.

COMPLEX MARTINGALE CONVERGENCE*

G. A. Edgar
The Ohio State University
Columbus, Ohio 43210

ABSTRACT

We investigate martingales appropriate for use in complex Banach spaces in connection with the complex uniform convexity popularized by Davis, Garling and Tomczak. This brings us into contact with diverse concepts, such as: pseudo-convex sets, plurisubharmonic functions, conformal martingales, the Radon-Nikodym property, and the analytic Radon-Nikodym property.

0. Introduction

Martingales with values in a Banach space have been used to study the structure of Banach spaces. One example is the equivalence of convergence of L^1-bounded martingales and the Radon-Nikodym property [7]. There are many other examples but convergence is the only one explicitly considered here.

When a Banach space has complex scalars (as opposed to real scalars), the techniques used for its study may have to take that into account. I will be concerned here with martingales that are useful in spaces over the complex numbers.

Here is a simple example of what is involved, taken from a paper by Davis, Garling and Tomczak [6]. Let $\Omega = [0,1[^{I}$, with product measure. If $\omega \in \Omega$, write $\omega_1, \omega_2, \ldots$ for its components. Define the sequence (X_n) of random variables by:

$$(*) \qquad X_n(\omega) = \sum_{k=1}^{n} f_k(\omega_1, \ldots, \omega_{k-1}) e^{2\pi i \omega_k}$$

(where the f_n's are measurable and bounded). Then (X_n) is a martingale of a very special form. If the f_n's have values in a Banach space E, then (X_n) is a martingale in E. For example, take $E = L^1([0,1])$. Then E fails Radon-Nikodym property. So there are L^1-bounded martingales in E that diverge. But in fact all L^1-bounded martingales in E

*Supported in part by N.S.F. grant DMS84-01986.

of the form (*) converge a.s. (see [2], Corollary 4.3). This paper arose as an attempt to understand this fact.

The martingales (*) form an interesting class, but they are too limited for many purposes. For example, this class is not closed under optional sampling: if $\tau_1 \le \tau_2 \le \tau_3 \le \ldots$ is a sequence of (bounded) stopping times, $Y_n = X_{\tau_n}$ is a new martingale. But if (X_n) is of the form (*), it does not follow that (Y_n) is of the form (*). In this paper I am concerned with how the definition should be extended to have more useful permanence properties, but still to retain the convergence properties exhibited by (*).

The concepts that are used include some from the field of several complex variables (as might be expected, especially when the Banach spaces are finite-dimensional). But care must be taken in how the concepts are formulated. Generally, derivatives must be avoided; many of the Banach spaces of interest do not admit equivalent differentiable norms. So, for example, there is no way to define "the unit ball is strictly pseudo-convex" using derivatives.

1. Preliminaries

We begin with the finite-dimensional definitions. Let U be an open set in $\mathbb{C}$. A function $\psi : U \to [-\infty, \infty[$ is called <u>subharmonic</u> on U iff ψ is upper semicontinuous and, for all $x \in U$ and all $y \in \mathbb{C}$, if $\{x + \lambda y : |\lambda| \le 1\} \subseteq U$, then

$$\int_0^{2\pi} \psi(x + e^{i\theta} y) \frac{d\theta}{2\pi} \ge \psi(x).$$

Note that we have allowed $\psi(x) \equiv -\infty$.

Let U be an open set in $\mathbb{C}^n$, where n is a positive integer. A function $\psi : U \to [-\infty, \infty[$ is called <u>plurisubharmonic</u> iff its restriction to each complex line in $\mathbb{C}^n$ is subharmonic. That is, if $x, y \in \mathbb{C}^n$, and $\theta : \mathbb{C} \to \mathbb{C}^n$ is defined by $\theta(\lambda) = x + \lambda y$, then $\psi \circ \theta$ is subharmonic on $\theta^{-1}[U]$.

Let U be an open set in $\mathbb{C}^n$. Then U is called <u>pseudoconvex</u> iff the function $\psi : U \to [-\infty, \infty[$ defined by

$$\psi(x) = -\log \operatorname{dist}(x, \mathbb{C}^n \setminus U)$$

is plurisubharmonic on U. We intend the distance in the Euclidean norm on $\mathbb{C}^n$, but an equivalent definition is obtained if any other norm on $\mathbb{C}^n$ is substituted.

Discussion of plurisubharmonic functions and pseudoconvex regions in $\mathbb{C}^n$ can be found in many text on several complex variables. One discussion, which emphasizes the similarity to convex functions and convex regions in $\mathbb{R}^n$, has been given by Bremermann [3].

It is natural to extend the definitions to infinite-dimensional spaces. (See, for example, [13]).

Let E be a complex topological vector space. Let U be an open set in E. Then U is said to be pseudoconvex iff $U \cap F$ is pseudoconvex for every finite-dimensional subspace F of E. A function $\psi : U \to [-\infty, \infty[$ is called plurisubharmonic iff ψ is upper semicontinuous and its restriction to each complex line in E is subharmonic. The last part can be rephrased as follows: If $x, y \in E$ and $\{x + \lambda y : |\lambda| \le 1\} \subseteq U$, then

$$\int_0^{2\pi} \psi(x + e^{i\theta} y) \frac{d\theta}{2\pi} \ge \psi(x).$$

A topological vector space is called locally pseudoconvex (or locally holomorphic [1]) iff there is a base of balanced pseudoconvex neighborhoods of the origin. (See also [1], [14].)

If U is an open balanced neighborhood of 0 and ϕ is its Minkowski gauge, then the following are equivalent (see [1, p.40]):

(a) U is pseudoconvex;
(b) ϕ is plurisubharmonic;
(c) $\log \phi$ is plurisubharmonic.

A quasi-norm on E is a function $\phi : E \to [0, \infty[$ satisfying: $\phi(\lambda x) = |\lambda| \phi(x)$ for $x \in E$, $\lambda \in \mathbb{C}$; $\phi(x + y) \le K(\phi(x) + \phi(y))$ for some constant K ; $\phi(x) = 0$ if and only if $x = 0$. For example, if (Ω, F, μ) is a measure space, and $0 < p < \infty$, then

$$\|f\| = \left(\int |f|^p \, d\mu\right)^{1/p}$$

defines a quasi-norm on $L^p(\mu)$. [It is a norm if $p \geq 1$.]

If E is a topological vector space whose topology is defined by a quasi-norm $\|\cdot\|$ that is uniformly continuous on bounded sets of E, then $(E, \|\cdot\|)$ is called a continuously quasi-normed space.

According to [6], a continuously quasi-normed space $(E, \|\cdot\|)$ is called locally PL-convex if and only if the function $\log\|\cdot\|$ is plurisubharmonic. Equivalently, the function $\|\cdot\|^p$ is plurisubharmonic for some (or all) p with $0 < p < \infty$, or the ball $\{x \in E : \|x\| < 1\}$ is pseudoconvex. Such a space is certainly locally pseudoconvex.

The most important class of examples for the present paper is the class of Banach spaces $(E, \|\cdot\|)$. Then $\|\cdot\|$ is a convex function, and therefore plurisubharmonic by Jensen's inequality.

A principle commonly used in (finite-dimensional) several complex variables says that a region is pseudoconvex if and only if it is "pseudoconvex at each boundary point". We state here an infinite-dimensional instance of this principle.

1.1 PROPOSITION. Let $(E, \|\cdot\|)$ be a continuously quasi-normed space. Then E is locally PL-convex if and only if, for every $x_o \in E$ with $\|x_o\| = 1$, there is a plurisubharmonic function ψ on E with $\psi(x_o) = 1$ and $\psi(x) \leq \|x\|$ for all x.

Proof. Suppose E is locally PL-convex, so that $\|\cdot\|$ is a plurisubharmonic function. Take $\psi(x) = \|x\|$. Conversely, suppose such functions ψ exist. We claim that $\|\cdot\|$ is plurisubharmonic. Let x_o, $y_o \in E$. If $x_o = 0$, then

$$\int_o^{2\pi} \|x_o + e^{i\theta} y_o\| \frac{d\theta}{2\pi} \geq 0 = \|x_o\|.$$

If $x_o \neq 0$, there is a plurisubharmonic function ψ with $\psi(x_o/\|x_o\|) = 1$

and $\psi(x) \leq \|x\|$ for all x, so

$$\int_0^{2\pi} \|x_o + e^{i\theta} y_o\| \frac{d\theta}{2\pi} = \|x_o\| \int_0^{2\pi} \| \frac{x_o}{\|x_o\|} + e^{i\theta} \frac{y_o}{\|x_o\|} \| \frac{d\theta}{2\pi}$$

$$\geq \|x_o\| \int_0^{2\pi} \psi(\frac{x_o}{\|x_o\|} + e^{i\theta} \frac{y_o}{\|x_o\|}) \frac{d\theta}{2\pi}$$

$$\geq \|x_o\| \psi(\frac{x_o}{\|x_o\|}) = \|x_o\|.$$

This shows that $\|\cdot\|$ is plurisubharmonic, and therefore that E is locally PL-convex.

The above Proposition may help explain the following definitions. Let $(E, \|\cdot\|)$ be a continuously quasi-normed space. Then $\|\cdot\|$ is <u>strictly plurisubharmonic</u> iff, for every $x_o \in E$ with $\|x_o\| = 1$ there exists a plurisubharmonic function ψ on E with $\psi(x_o) = 1$ and $\psi(x) < \|x\|$ for all $x \neq x_o$. Similarly, $\|\cdot\|$ is <u>uniformly plurisubharmonic</u> iff there is a continuous, increasing function $h : [0,\infty[\to [0,\infty[$, with $h(0) = 0$ and $h(t) > 0$ for $t > 0$, such that for every $x_o \in E$ with $\|x_o\| = 1$, there exists a plurisubharmonic function ψ on E with $\psi(x_o) = 1$ and $\psi(x) \leq \|x\| - h(\|x - x_o\|)$ for all $x \in E$.

The quasi-norm $(\int \|f\|^p d\mu)^{1/p}$ on $L^p(\mu)$ is uniformly plurisubharmonic, if $0 < p < \infty$. (See Section 3, and compare with [6] and [14].)

The reader may find it instructive to investigate whether replacing "plurisubharmonic" by "convex" yields conditions equivalent to strict convexity and uniform convexity.

The definitions above were inspired by a definition of Davis-Garling-Tomczak [6]. They say that E is <u>uniformly PL-convex</u> iff there is a function $h : [0,\infty[\to [0,\infty[$ with $h(t) > 0$ for $t > 0$, such that for all $x_o \in E$

with $\|x_o\| = 1$, and all $y \in E$,

$$\int_0^{2\pi} \|x_o + e^{i\theta}y\| \frac{d\theta}{2\pi} \geq 1 + h(\|y\|).$$

It is easy to see that this is true if $\|\cdot\|$ is uniformly plurisubharmonic. I do not know whether the converse is true; I expect that it is not.

The goal of this paper is to discuss convergence of martingales. The class of martingales must be restricted (to obtain convergence in L^1, for example). There are several possibilities for definitions; I have chosen one of them to use there. For simplicity, the discussion will be restricted primarily to the case of a separable Banach space E. We will write PSH(E) for the set of all plurisubharmonic functions on E. Similarly, PSH(V) is the set of all plurisubharmonic function on the open set V.

Let μ, υ be probability measures on the Borel subsets of E with first moment (i.e., $\int \|x\| d\mu(x) < \infty$, etc.). We say μ <u>dominates</u> υ, and write $\mu > \upsilon$ [PSH(E)], iff

$$\int \psi d\mu \geq \int \psi d\upsilon$$

for all $\psi \in$ PSH(E). For $x \in E$, we say μ is a <u>Jensen measure</u> for x, and write $\mu \sim x$[PSH(E)] iff μ dominates the Dirac measure ε_x, that is,

$$\int \psi d\mu \geq \psi(x)$$

for all $\psi \in$ PSH(E). (See Gamelin [8] for discussion of Jensen measures in the finite-dimensional case.)

One good example of $\mu \sim x$[PSH(E)] is obtained as follows. If $x, y \in E$, then the uniform distribution μ on the circle $x + e^{i\theta}y$, $0 \leq \theta < 2\pi$, represents x. That is,

$$\int_0^{2\pi} \psi(x + e^{i\theta}y) \frac{d\theta}{2\pi} \geq \psi(x)$$

for all $\psi \in PSH(E)$. Another example can be obtained as follows. Let υ be a probability measure in $\mathbb{C}$ and $\upsilon \sim 0$ [PSH($\mathbb{C}$)]. Suppose $f : \mathbb{C} \to E$ is holomorphic. Then the image measure $\mu = f(\upsilon)$ represents $f(0)$. The reason for this is that for any plurisubharmonic function ψ on E, the composition $\psi \circ f$ is subharmonic on $\mathbb{C}$.

If μ is a Jensen measure for x, then of course the barycenter of μ is x; i.e., $\int y\, \mu(dy) = x$. (If E is a separable Banach space, this exists as a Bochner integral.) The reason is that the real and imaginary parts of a linear functional are pluriharmonic, so we get

$$\int f(y)\,\mu(dy) = f(x)$$

for all $f \in E^*$. In general it is not enough to use only pluriharmonic functions in the definition of Jensen measures. Here is an example in one complex dimension, from [8]. The Poisson kernel

$$P_r(\theta) = \frac{1 - r^2}{1-2r\cos\theta+r^2}$$

satisfies

$$P_0(\theta) = 1,$$

$$\min P_{1/2}(\theta) = \frac{1}{3},$$

$$\max P_{1/2}(\theta) = 3.$$

Let the measure μ on $\mathbb{C}$ be defined by

$$\mu = \frac{1}{3}\,\varepsilon_{1/2} + [1 - \frac{1}{3}\,P_{1/2}(\theta)]\lambda\ ,$$

where $d\lambda = d\theta/2\pi$ is Haar measure on the unit circle. Then μ is a probability measure. If f is harmonic on the closed disk, then

$$\int f d\mu = \frac{1}{3} f(\tfrac{1}{2}) + \int_0^{2\pi} f(e^{i\theta})[1 - \tfrac{1}{3}P_{1/2}(\theta)]\,\frac{d\theta}{2\pi}$$

$$= \frac{1}{3}\,f(\tfrac{1}{2}) + f(0) - \frac{1}{3}\,f(\tfrac{1}{2}) = f(0).$$

So μ represents 0 for all <u>harmonic</u> f. (In the language of Gamelin

[8], μ is an "Arens-Singer measure" for 0.) But μ is not a Jensen measure for 0. The function $\log|z|$ is subharmonic, but

$$\log|\tfrac{1}{2}| = \log|\tfrac{1}{2} - 0| \not\leq \int \log|\tfrac{1}{2} - z|\mu(dz) = -\infty.$$

Finally, we turn to the definitions concerning martingales. A discussion of real-scalar martingales in a Banach space can be found in [7, Chapter V].

Let (Ω, F, P) be a probability space. We will write $E[X] = \int X dP$ for the expectation if X is a random variable. Let $(F_n)_{n=0}^{\infty}$ be an increasing sequence of σ-algebras contained in F, and let E be a separable Banach space. If X_n is an F_n-Bochner integrable random variable, $X_n : \Omega \to E(n = 0,1,2,\ldots)$, then we will say that the sequence (X_n) is a PSH(E)-<u>martingale</u> iff, for every $\psi \in$ PSH(E), the real-valued process $(\psi(X_n))_{n=0}^{\infty}$ is a submartingale. Since the real part of a linear functional is plurisubharmonic, a PSH(E)-martingale is, in particular, a martingale in the usual sense. Another way to think of this is the following: Given X_n, the distribution of X_{n+1} is a measure that dominates the point X_n in the sense defined above. It follows that the image measures increase: $X_0(P) < X_1(P) < X_2(P) < \ldots$ [PSH(E)].

A useful class of examples can be found in [6]. Suppose $v_n : \Omega \to E$ is F_{n-1}-measurable $(n > 0)$ and v_0 is constant in E; η_n is uniformly distributed on $\{\lambda \in \mathbb{C} : |\lambda| = 1\}$, η_n is independent of F_{n-1}, η_n is F_n-measurable. Then if $v_n \in L^p(\mu, E)$, the process

$$X_n = v_0 + \sum_{k=1}^{n} \eta_k v_k$$

is called an H_p-<u>shrub</u> in [6] and called an <u>analytic martingale</u> in [2]. It can be seen that analytic martingales are PSH(E)-martingales. (The H_p-martingales defined in [6] are not used in this paper, since they need not converge, even in L^1.)

There is a connection with the <u>conformal martingales</u> of Getoor and Shape

[10]. If (X_t) is a martingale in E with continuous trajectories, then according to Proposition 5.9 of [16], (X_t) is a conformal martingale if and only if $(\psi(X_t))$ is a submartingale for any plurisubharmonic function where it makes sense. (Schwartz says these results "ne sont sans doute pas plus que des amusements".) But for discrete parameter martingales the situation is more complicated. In $\mathbb{C}$, there are martingales (X_n) such that (X_n^2) is also a martingale, but (X_n^3) is not a martingale. For example, $X_0 = 0$, but $X_1 = X_2 = X_3 \ldots$ and

$$P[X_1 = 1] = \frac{1}{4}$$

$$P[X_1 = -1] = \frac{1}{4}$$

$$P[X_1 = i] = \frac{1}{4}$$

$$P[X_1 = -i] = \frac{1}{4}.$$

2. Martingale convergence

In this section it is proved that PSH(E)-martingales, bounded in $L^1(E)$-norm, converge a.s., provided the norm in E is uniformly plurisubharmonic. Some remarks on relaxing this condition are included in Section 4.

Throughout this section, $(E, \|\cdot\|)$ is a separable Banach space. Many of the assertions also hold in separable continuously quasi-normed spaces, or even non-separable spaces; but I will not spell that out here.

2.1 LEMMA. Suppose $\phi : E \to [-\infty, \infty[$ is upper semicontinuous and bounded above on bounded sets. Define $\psi_0(x) = \phi(x)$ and

$$\psi_{n+1}(x) = \inf \{\int_0^{2\pi} \psi_n(x + e^{i\theta}v) \frac{d\theta}{2\pi} : v \in E\}$$

for $n \geq 0$. Then ψ_n decreases pointwise to the largest plurisubharmonic function less than or equal to ϕ.

Proof. Taking $v = 0$, we see that $\psi_{n+1}(x) \leq \psi_n(x)$, so $\psi(x) = \lim_{n\to\infty} \psi_n(x)$

exists for each $x \in E$. We will prove by induction that each ψ_n is upper semicontinuous. First, $\psi_0 = \phi$ is upper semicontinuous. Suppose ψ_n is upper semicontinuous, and consider ψ_{n+1}. For fixed x,v the map $\theta \to \psi_n(x + e^{i\theta} v)$ is bounded above and upper semicontinuous, hence measurable. For fixed v, I claim that $x \to \int_0^{2\pi} \psi_n(x + e^{i\theta} v) \frac{d\theta}{2\pi}$ is upper semicontinuous. Indeed, if $x_k \to x$, then (by the upper semicontinuity of ψ_n) we have $\limsup_{k\to\infty} \psi_n(x_k + e^{i\theta} v) \leq \psi_n(x + e^{i\theta} v)$ for all θ. Now this is bounded above, so by Fatou's Lemma

$$\limsup_{k\to\infty} \int_0^{2\pi} \psi_n(x_k + e^{i\theta} v) \frac{d\theta}{2\pi} \leq \int_0^{2\pi} \limsup_{k\to\infty} \psi_n(x_k + e^{i\theta} v) \frac{d\theta}{2\pi}$$

$$\leq \int_0^{2\pi} \psi_n(x + e^{i\theta} v) \frac{d\theta}{2\pi} .$$

This shows that $x \to \int_0^{2\pi} \psi_n(x + e^{i\theta} v) \frac{d\theta}{2\pi}$ is upper semicontinuous. So ψ_{n+1} is the infimum of a family of upper semicontinuous functions, so it is upper semicontinuous. This completes the induction.

We next show that the limit ψ is plurisubharmonic. Fix x,v. Then by the monotone convergence theorem

$$\int_0^{2\pi} \psi(x + e^{i\theta} v) \frac{d\theta}{2\pi} = \lim_n \int_0^{2\pi} \psi_n(x + e^{i\theta} v) \frac{d\theta}{2\pi}$$

$$\geq \lim_n \psi_{n+1}(x) = \psi(x) .$$

So ψ is plurisubharmonic.

Next we show that ψ is the largest plurisubharmonic function $\leq \phi$. Suppose ψ' is any plurisubharmonic function $\leq \phi$. Then $\psi' \leq \psi_0 = \phi$. By induction we see that $\psi' \leq \psi_n$ for all n, so $\psi' \leq \psi$.

The construction in this lemma can be rephrased in terms of PSH(E)-martingales. Suppose E is a separable Banach space, and ϕ, ψ_n, ψ are as in the Lemma. Then I claim that

$$\psi_n(x) = \inf E[\phi(x_n)] ,$$

where the infimum is over all analytic martingales $(X_k)_{k=0}^n$ with $X_o = x$. The proof is by induction on n. For $n=0$, we have only $X_o = x$, so $\psi_o(x) = \phi(x) = E[\phi(X_o)]$. Suppose the formula is known for $n-1$. Fix $x_o \in E$ and $\varepsilon > 0$. Choose v so that

$$\psi_n(x_o) \geq \int_0^{2\pi} \psi_{n-1}(x_o + e^{i\theta} v) \frac{d\theta}{2\pi} + \frac{\varepsilon}{2} .$$

Let X_1 have the uniform distribution on the circle $x_o + e^{i\theta} v$. Now choose measurably (by the Yankov-von Neumann selection theorem [12]) for each θ an analytic martingale $(X_k)_{k=1}^n$ with $X_1 = x_o + e^{i\theta} v$ and

$$\psi_{n-1}(x_o + e^{i\theta} v) \geq E[\phi(X_n)] + \frac{\varepsilon}{2} .$$

Thus we have obtained $(X_k)_{k=1}^n$ conditionally on X_1. Putting them together, we get $(X)_{k=0}^n$ with $X_o = x_o$, $E[\phi(X_n)] + \varepsilon \leq \psi_n(x_o)$.

2.2 PROPOSITION. Let E be a separable complex Banach space; let $0 < p < \infty$; let $h : [0, \infty[\to [0, \infty[$ be increasing and continuous. Then the following are equivalent:

(a) For every $x_o \in E$ with $\|x_o\| = 1$, there is a plurisubharmonic $\psi : E \to [-\infty, \infty[$ with $\psi(x_o) = 0$ and for all $x \in E$,

$$\psi(x) \leq \|x\|^p - 1 - h(\|x - x_o\|) .$$

(b) For every $x_o \in E$ with $\|x_o\| = 1$, and every analytic martingale (X_n) with $X_o = x_o$, we have

$$E[\|X_n\|^p] \geq 1 + E[h(\|X_n - x_o\|)] .$$

(c) For every $x_o \in E$ with $\|x_o\| = 1$, and every Borel measure γ on E representing x_o, we have

$$\int \|x\|^p \, d\gamma(x) \geq 1 + \int h(\|x - x_o\|) \, d\gamma(x) \; .$$

Proof. (c) $\Rightarrow$ (b). If (X_n) is an H_∞-shrub with $X_o = x_o$, then the distribution γ of X_n represents x_o.

(a) $\Rightarrow$ (c). Since ψ is plurisubharmonic, we have $\int \psi d\gamma \geq \psi(x_o)$ if $\gamma \sim x_o$ [PSH(E)]. Thus

$$\int \|x\|^p \, d\gamma(x) \geq \int (\psi(x) + 1 + h(\|x - x_o\|)) \, d\gamma(x)$$

$$\geq 0 + 1 + \int h(\|x - x_o\|) \, d\gamma(x).$$

(b) $\Rightarrow$ (a). Fix $x_o \in$ with $\|x_o\| = 1$. Define a function ψ_o by $\psi_o(x) = \|x\|^p - 1 - h(\|x - x_o\|)$. Then ψ_o is continuous and bounded on bounded sets. Let ψ be the largest plurisubharmonic function $\leq \psi_o$. It remains only to show $\psi(x_o) = 0$. Now $\psi = \lim_n \psi_n$, where ψ is as in Lemma 2.1. Since $\psi_n(x_o) = 0$, it suffices to show $\psi_n(x_o) \geq 0$ for all n. But $\psi_n(x_o) = \inf E\,[\psi_o(X_n)]$, where the infimum is over all analytic martingales $(X_k)_{k=0}^n$ with $X_o = x_o$. By (b), we have $E[\psi_o(X_n)] = E[\|X_n\|^p - 1 - h(\|X_n - x_o\|)] \geq 0$.

According to this proposition, $\|\cdot\|$ is uniformly plurisubharmonic if and only if there is an h with $h(t) > 0$ for which (a), (b), (c) hold with p=1. It may be useful to note that "uniform PL-convexity" of [6] can be rephrased as follows: There is $h:[0,\infty[\to[0,\infty[$, increasing, continuous, $h(t) > 0$ for $t > 0$, such that for every $x_o \in E$, $\|x_o\| = 1$, and every analytic martingale $(X_k)_{k=0}^1$ with $X_o = x_o$, we have $E[\|X_1\|^p] \geq 1 + E[h(\|X_1 - x_o\|)]$. In fact, if this holds for one value of p $(0<p<\infty)$, then it holds for all.

The conditions in Proposition 2.2 are not changed if an exponent 1/p is added, as in the following.

2.3 PROPOSITION. Let $(E, \|\cdot\|)$ be a continuously quasi-normed space, and

$0 < p < \infty$. Then the following are equivalent:

(a) There is a function $h : [0,\infty[\to[0,\infty[$, increasing, continuous, $h(t) > 0$ for $t > 0$, such that for all $x_0 \in E$ with $\|x_0\| = 1$ and all measures $\gamma \sim x_0$,

$$\int \|x\|^p d\gamma(x) \geq 1 + \int h(\|x - x_0\|)\, d\gamma(x) .$$

(b) There is a function $k : [0,\infty[\to [0,\infty[$, increasing, continuous, $k(t) > 0$ for $t > 0$, such that for all $x_0 \in E$ with $\|x_0\| = 1$ and all measures $\gamma \sim x_0$,

$$\left(\int \|x\|^p\, d\gamma(x)\right)^{1/p} \geq 1 + \int k(\|x - x_0\|)\, d\gamma(x) .$$

<u>Proof</u>. If $p \geq 1$ and (a) holds, let $k(t) = (1 + h(t))^{1/p} - 1$. If $p \geq 1$ and (b) holds, let $h(t) = pk(t)$. If $p < 1$ and (a) holds, let $k(t) = ph(t)$. If $p < 1$ and (b) holds, let $h(t) = (1 + k(t))^p - 1$. The verifications, involving Jensen's inequality and Bernoulli's inequality, are omitted.

The following martingale convergence theorem is now easy to prove.

<u>2.4 THEOREM</u>. Let $(E, \|\cdot\|)$ be a Banach space; suppose $\|\cdot\|$ is uniformly plurisubharmonic. Let (X_n) be a PSH(E)-martingale; suppose $\sup E[\|X_n\|] < \infty$. Then X_n converges a.s.

<u>Proof</u>. Since all of the X_n are Bochner integrable, we may assume E is separable. We begin with (c) of Proposition 2.2: There is a function h with $h(t) > 0$ for $t > 0$ such that if $\|x_0\| = 1$ and $\gamma \sim x_0$ [PSH(E)], then

$$\int \|x\| d\gamma(x) \geq 1 + \int h(\|x - x_0\|) d\gamma(x) .$$

From this follows, for $x_0 \in E$, $x_0 \neq 0$, that if $\gamma \sim x_0$[PSH(E)], then

$$\int \|x\| d\gamma(x) \geq \|x_0\| + \|x_0\| \int h\left(\frac{\|x - x_0\|}{\|x_0\|}\right) d\gamma(x) .$$

Thus, if (X_n) is a PSH(E)-martingale, with $X_o = x_o$, we have

$$E[\|X_n\|] \geq \|x_o\| + \|x_o\| E[h(\frac{\|X_n - x_o\|}{\|x_o\|})] .$$

The conditional version of this implies that if (X_n) is a PSH(E)-martingale with respect to the σ-algebras (F_n), and if $n > m$, then

$$E[\|X_n\| \mid F_m] \geq \|X_m\| + \|X_m\| E[h(\frac{\|X_n - X_m\|}{\|X_m\|}) \mid F_m] .$$

Integrating this, we get

$$E[\|X_m\|] \geq E[\|X_m\|] + E[\|X_m\| h(\frac{\|X_n - X_m\|}{\|X_m\|})] .$$

Now $(X_n)_{n=0}^{\infty}$ is a PSH(E)-martingale, and $\|\cdot\|$ is plurisubharmonic, so $(\|X_n\|)_{n=0}^{\infty}$ is a submartingale, so it converges a.s., and $E[\|X_n\|]$ increases. Thus, as n and m increase without bound, we see that

$$E[\|X_n\|] - E[\|X_m\|] \to 0$$

so that

$$E[\|X_n\| \ h(\frac{\|X_n - X_m\|}{\|X_m\|})] \to 0 .$$

Thus $\|X_n\| \ h(\frac{\|X_n - X_m\|}{\|X_m\|})$ converges to zero in probability. But $\|X_n\|$ converges a.s., so $h(\frac{\|X_n - X_m\|}{\|X_m\|})$ converges to zero in probability (on the set where $\|X_n\|$ does not converge to zero), so $\|X_n - X_m\|$ converges to zero in probability. But (X_n) is a martingale, so in fact it converges a.e.

3. Integrable function spaces.

An interesting example of a space with uniformly plurisubharmonic norm is the space $L^1(\mu)$, where (Ω, F, μ) is a measure space. The first case is the one-dimensional space $\mathbb{C}$. The complex number 1 is the typical point on the surface of the unit ball. The function

$$h(t) = \frac{1}{16}\,\frac{t^2}{1+t}$$

satisfies $h(t) > 0$ for $t > 0$. The function

$$\psi(z) = \frac{1}{2}(\log|z| + \operatorname{Re} z - 1)$$

is subharmonic on $\mathbb{C}$. It is an elementary exercise to verify that

$$\psi(z) \leq |z| - 1 - h(|z - 1|)$$

for all $z \in \mathbb{C}$. So by Proposition 2.2, the absolute value $|\cdot|$ is "uniformly subharmonic".

So, if γ is a probability measure on $\mathbb{C}$ with $\gamma \sim 1$ [PSH($\mathbb{C}$)], we have

$$\int |z|\,d\gamma(z) \geq 1 + \int h(|z - 1|)\,d\gamma(z).$$

More generally, if z_0 is any complex number, and $\gamma \sim z_0$ [PSH($\mathbb{C}$)], then

$$(*) \qquad \int |z|\;d\gamma(z) \geq |z_0| + |z_0| \int h\left(\frac{|z - z_0|}{|z_0|}\right)d\gamma(z).$$

In order to extend to function spaces $L^1(\mu)$, there are two possibilities. One is to start with $f_0 \in L^1(\mu)$, $\|f_0\| = 1$, and reduce to the case $f_0 > 0$. Then

$$\psi(f) = \frac{1}{2}\int [f_0(\log|f| - \log|f_0|) + \operatorname{Re} f - \operatorname{Re} f_0]\,d\mu$$

is plurisubharmonic on $L^1(\mu)$ and satisfies

$$\psi(f) \leq ||f|| - 1 - h(||f - f_o||) .$$

The other possibility is to start with γ on $L^1(\mu)$, $\gamma \sim f_o$, then "disintegrate" it as $(\gamma_\omega)_{\omega \in \Omega}$, where $\gamma_\omega \sim f_o(\omega)$ for almost all ω. Then integrate the inequality (*) over Ω. The details are omitted.

4. Is there a complex Radon-Nikodym property?

For real Banach spaces, uniform convexity implies a martingale convergence theorem. But uniform convexity is a very strong condition; the condition that is both necessary and sufficient for martingale convergence (the Radon-Nikodym property) is interesting in its own right. This section contains some suggestions for a corresponding complex notion. However, the question is not answered here.

Let E be a complex Banach space. Under what conditions does every L^1-bounded PSH(E)-martingale converge? Is there a criterion like dentability (cf. [7, p. 133]), or a vector measure criterion (cf. [7, p. 127])? If C is a bounded subset of E, under what conditions does every PSH(E)-martingale with values in C converge? If C is open, it might be reasonable to consider PSH(C)-martingales, that is, sequences (X_n) such that $(\psi(X_n))$ is a submartingale for every plurisubharmonic function ψ defined on C.

Representability in the sense used here ($\mu \sim x$[PSH(E)]) is more difficult to manage than the real version. This is illustrated in the following. The notation C_ε, where C is a set and $\varepsilon > 0$, will signify the ε-neighborhood $\{x \in E : \text{dist}(x,C) < \varepsilon\}$ of C.

<u>4.1 PROPOSITION</u>. Let C be a Borel subset of the separable Banach space E, and let $x_o \in C$. Then (1) $\Rightarrow$ (2) $\Rightarrow$ (3) $\Leftrightarrow$ (4) $\Rightarrow$ (5):

(1) There is a probability measure μ on C with $\mu \sim x_o$ [PSH(E)];

(2) There is no function $\psi \in$ PSH(E) with $\psi \leq 0$ on C but $\psi(x_o) > 0$;

(3) There is no function $\psi \in$ PSH(E) with $\psi(x) \leq \text{dist}(x,C)$ for all $x \in E$ and $\psi(x_o) > 0$;

(4) For any $\varepsilon > 0$, there is no analytic martingale $(X_k)_{k=0}^n$ with $X_o = x_o$ and $E[\text{dist}(X_n,C)] < \varepsilon$;

(5) For any $\varepsilon, \varepsilon' > 0$, there is a probability measure $\mu \sim x_o[PSH(E)]$ with $\mu(C_\varepsilon) > 1 - \varepsilon'$.

Proof. (1) => (2). If $\psi \in PSH(E)$ and $\psi < 0$ on C, then $\psi(x_o) \leq \int_E \psi(x)\, d\mu(x) = \int_C \psi(x)\, d\mu(x) \leq 0$.

(2) => (3). If $x \in C$, then $\text{dist}(x,C) = 0$.

(3) => (4). Let $\psi_o(x) = \text{dist}(x,C)$. Let ψ be the largest plurisubharmonic function $\leq \psi_o$. By assumption, $\psi(x_o) \leq 0$. If the sequence ψ_n is defined as in Lemma 2.1, then there is n so that $\psi_n(x_o) < \varepsilon$. Therefore, there is an analytic martingale $(X_k)_{k=o}^n$ with $X_o = x_o$ and $E[\psi_o(X_n)] < \varepsilon$.

(4) => (3) is similar.

(4) => (5). Choose an analytic martingale $(X_k)_{k=o}^n$ with $X_o = x_o$ and $E[\text{dist}(X_n,C)] < \varepsilon\varepsilon'$. Let μ be the distribution of X_n, so $\mu(A) = P[X_n \in A]$ for Borel sets A. Then $\mu(C_\varepsilon) = P[d(X_n,C) < \varepsilon] > 1 - \varepsilon'$.

I do not know whether the conditions are equivalent under reasonable circumstances (such as C compact or C convex).

Martingale convergence leads to properties that resemble dentability. Here is one of the simplest such properties. (If "plurisubharmonic" is replaced by "convex", the conclusion would be that U is dentable.)

4.2 PROPOSITION. Let E be a separable Banach space. Suppose every L^1-bounded PSH(E)-martingale converges a.s. Let U be the open unit ball in E, and let $\varepsilon > 0$. Then there is a plurisubharmonic function ψ so that the set $\{x \in U : \psi(x) > 0\}$ is nonempty and has diameter less than ε.

Proof. Suppose there is $\varepsilon > 0$ so that if ψ is any plurisubharmonic function and $\psi(x) > 0$ somewhere on U, then diam $\{x \in U : \psi(x) > 0\} > \varepsilon$. We will construct a nonconvergent PSH(E)-martingale.

The probability space will be $\Omega = [0,1]^{\mathbb{N}}$ with P the product Lebesgue measure, and F_n the σ-algebra determined by the first n coordinates in Ω. (In fact $[0,1]$ maps measurably onto any complete separable metric space, such as E, so this choice is unimportant.) Let $a_n = 1 - 2^{-n}$. There will be constructed sets $\Omega_n \in F_n$ with $P[\Omega_n] = 2^{-1} + 2^{-(n+1)}$, and random variables X_n with $E[\mathrm{dist}(X_n, a_{n-1}U)1_{\Omega_n}] < 2^{-(2n+2)}$, and $P\{\omega \in \Omega_n : \|X_n(\omega) - X_{n-1}(\omega)\| < \varepsilon/8\} < \varepsilon^{-1}2^{-(2n-1)}$.

Let $X_o = 0$, $\Omega_o = \Omega$. Suppose X_n has been constructed with $E[\mathrm{dist}(X_n, a_{n-1}U)1_{\Omega_n}] < 2^{-(2n+2)}$. Then $P\{\omega \in \Omega_n : \|X_n(\omega)\| > a_n\} < 2^{-(n+2)}$ since $a_n = a_{n-1} + 2^{-n}$. Choose $\Omega_{n+1} \subseteq \Omega_n \cap \{\|X_n\| \le a_n\}$ with $P[\Omega_{n+1}] = 2^{-1} + 2^{-(n+2)}$. On $\Omega \setminus \Omega_{n+1}$, let $X_{n+1} = X_n$. On Ω_{n+1}, proceed as follows. Suppose $X_n(\omega) = x$, where $\|x\| \le a_n$. Let $D = \{Y: \|y - x\| < \varepsilon/4\}$, so that $\mathrm{diam}(a_nU \cap D) < 2\ \varepsilon/4$, and hence $\mathrm{diam}\ (U \cap a_n^{-1}D) < \varepsilon a_n^{-1}/2 \le \varepsilon$. Thus if $\psi \in PSH(E)$ is ≤ 0 on $a_nU \setminus D$, then $\psi \le 0$ on a_nU, so $\psi(x) \le 0$. By Proposition 5, (2) = (4), there is a random variable Y representing x $[PSH(E)]$ with $E[\mathrm{dist}(Y, a_nU \setminus D)] < 2^{-(2n+4)}$. Thus $E[\mathrm{dist}(Y, a_nU)] < 2^{-(2n+4)}$ and $P[\|Y - x\| < \varepsilon/8] < (8/\varepsilon)2^{-(2n+4)} = 2^{-(2n-1)}/\varepsilon$. The next step X_{n+1} will be chosen so that the conditional distribution, given $X_n = x$, is the distribution of Y. (These Y's should be chosen to depend measurably on x using the Yankov-von Neumann selection theorem.) So we get X_{n+1} with $X_{n+1} = X_n$ on $\Omega \setminus \Omega_{n+1}$, and $E[\mathrm{dist}(X_{n+1}, a_nU)1_{\Omega_{n+1}}] < 2^{-(2n+4)}$ and $P\{\omega \in \Omega_{n+1} : \|X_{n+1}(\omega) - X_n(\omega)\| < \varepsilon/8\} < 2^{-(2n-1)}/\varepsilon$. This completes the recursive construction of (X_n).

Now the $PSH(E)$-martingale (X_n) is L^1-bounded, since

$$E[\|X_n\|] = E[\sum_{k=0}^{n-1} \|X_k\| 1_{\Omega_k \setminus \Omega_{k+1}} + \|X_n\| 1_{\Omega_n}]$$

$$\le 1 + \sum_{k=1}^{n} E[\mathrm{dist}(X_k, U)1_{\Omega_k}]$$

$$\le 1 + \sum_{k=1}^{n} 2^{-(2k+2)} < 2.$$

Let $\Omega_\infty = \bigcap_{n=1}^{\infty} \Omega_n$. Then $P[\Omega_\infty] = 1/2$, but the series $\sum_{n=1}^{\infty} 2^{-(2n-2)}/\varepsilon$ converges, so a.s. on Ω_∞ we have $\|X_{n+1}(\omega) - X_n(\omega)\| \geq \varepsilon/8$ for all but finitely many n. Thus (X_n) does not converge a.s.

Notice that the martingale convergence is an isomorphic property, so the conclusion will hold also if U is the unit ball for an equivalent norm. I think it is unlikely that the converse is true. Another condition like dentability is given in the next result.

<u>4.3 PROPOSITION</u>. Suppose every L^1-bounded PSH(E)-martingale converges a.s. Let $\phi \in \text{PSH}(E)$ satisfy $\phi(X) \leq a\|X\| + b$, and let $\alpha \in \mathbb{R}$ satisfy $\inf \phi < \alpha < \sup \phi$. Then for any $\varepsilon > 0$, there is $\psi \in \text{PSH}(E)$ so that diam $\{x : \psi(x) > \phi(x)\} < \varepsilon$ and $\{x : \psi(x) > \phi(x), \alpha > \phi(x)\} \neq \emptyset$.

<u>Proof</u>. Suppose the conclusion is false. Then there is $\varepsilon > 0$ such that for all $\psi \in \text{PSH}(E)$, and all $x_0 \in E$ such that $\phi(x_0) < \alpha$, if $\psi \leq \phi$ on $\{x : \|x - x_0\| > \varepsilon/2\}$ then $\psi \leq \phi$ everywhere.

Fix x_0 with $\phi(x_0) < \alpha$, and let $\delta > 0$ be so small that $\phi(x_0) + \delta < \alpha$. Let $M = 1 + \sup\{\phi(x) : \|x - x_0\| < \varepsilon/2\}$. Define

$$\psi_0(x) = \begin{cases} M, & \text{if } \|x - x_0\| \leq \varepsilon/2 \\ \phi(x), & \text{if } \|x - x_0\| > \varepsilon/2 . \end{cases}$$

Now ϕ is upper semicontinuous and $M > \sup\{\phi(x) : \|x - x_0\| \leq \varepsilon/2\}$, it follows that ψ_0 is upper semicontinuous. Define ψ_n as in Lemma 2.1, so that ψ_n decreases to ψ, the largest plurisubharmonic function $\leq \psi_0$. Now $\psi \leq \phi$ on $\{x : \|x - x_0\| > \varepsilon/2\}$, so $\psi \leq \phi$ everywhere. In particular, $\psi(x_0) \leq \phi(x_0)$. Now there is n so that $\psi_n(x_0) < \phi(x_0) + \delta$, so there is an anlytic martingale $(X_k)_{k=0}^{n}$ with $X_0 = x_0$ and $E[\psi_0(X_n)] < \phi(x_0) + \delta$. Thus,

$$E[1_{\{\|X_n - x_0\| > \varepsilon/2\}} \phi(X_n)] + MP[\|X_n - x_n\| \leq \varepsilon/2] < \phi(x_0) + \delta .$$

Since $M = 1 + \sup\{\phi(x) : \|x - x_0\| \leq \varepsilon/2\}$, we get

$$E[1_{\{\|X_n - x_o\| > \varepsilon/2\}}\phi(X_n)] + E[1_{\{\|X_n - x_o\| \le \varepsilon/2\}}(X_n)] + P[\|X_n - x_o\| < \varepsilon/2]$$

$$< \phi(x_o) + \delta ,$$

so that

$$E[\phi(X_n)] + P[\|X_n - x_o\| \le \varepsilon/2] < \phi(x_o) + \delta .$$

But ϕ is plurisubharmonic, so $E[\phi(X_n)] \geq \phi(x_o)$. Thus $P[\|X_n - x_o\| \le \varepsilon/2] < \delta$ and $E[\phi(X_n)] - \phi(x_o) < \delta$. Therefore $E[\phi_n(X)] < \alpha$.

Using this construction, we will construct a martingale (Y_n) as before. Start at some point x_o with $\phi(x_o) < \alpha$. Choose δ_n decreasing rapidly to 0. We will get $\lim_n E[\phi(Y_n)] < \alpha$. If Y_n is defined, stop if $\phi(Y_n) \geq \alpha$, otherwise use the above to get Y_{n+1} with

$$P[\|Y_{n+1} - Y_n\| > \varepsilon/2 | \phi(Y_n) < \alpha] > 1 - \delta_n .$$

If δ_n converges to 0 fast enough, then $\lim_n E[\phi(Y_n)] < \alpha$, so (Y_n) is L^1-bounded, and we have $P[\lim_n \phi(Y_n) < \alpha] > 0$, and (Y_n) does not converge there.

This result, and others like it, may be more useful when stated in terms of functions in PSH(U) and PSH(U)-martingales, for some open set U, so that the martingale can be constructed inside the set U.

Here is one further remark. Fix $\eta > 0$, and U an open bounded set. If (X_n) is a PSH(U)-martingale, let

$$Q((X_n),\eta) = \{\omega : \|X_{n+1}(\omega) - X_n(\omega)\| < \eta \text{ except for finitely many } n\}.$$

Then let $\phi_\eta(x) = \inf P[Q((X_n),\eta)]$, where the inf is over all PSH(U)-martingales with $X_o = x$. The function ϕ_η is formally plurisubharmonic on U. Convergence of PSH(U)-martingales is related to whether $\phi_\eta(x)=1$ for all $\eta > 0$.

The paper [5] of Bukhvalov and Danilevich has recently been pointed out to me. It is concerned with the "analytic Radon-Nikodym property" for a complex Banach space E. It is characterized by the existence of boundary values for E-valued H^p-functions. It would be interesting to know whether there is a connection between the analytic Radon-Nikodym property and topics of this paper.

REFERENCES

[1] A.B. Aleksandrov, Essays on nonlocally convex Hardy classes. Lecture Notes in Math. 864, Springer-Verlag 1981, pp. 1-89.

[2] J. Bourgain and W.J. Davis, Martingale transforms and complex uniform convexity. Preprint.

[3] H.J. Bremermann, Complex convexity. Trans. Amer. Math. Soc. 82(1956) 17-51.

[4] ______________, Holomorphic functionals and complex convexity in Banach spaces. Pacific J. Math. 7(1957)811-831.

[5] A.V. Bukhvalov and A.A. Danilevich, Boundary properties of analytic and harmonic functions with values in Banach space. Mat. Zametki 31, No. 2 (1982)203-214. English translation: Math. Notes 31 (1982) 104-110.

[6] W.J. Davis, D.J.H. Garling and N. Tomczak-Jaegermann, The complex convexity of quasi-normed linear spaces. J. Funct. Anal. 55(1984) 110-150.

[7] J. Diestel and J.J. Uhl, Vector measures. Mathematical Surveys 15. American Mathematical Society 1977.

[8] T. W. Gamelin, Uniform algebras and Jensen measures. Cambridge University Press, 1978.

[9] D.J.H. Garling and N. Tomczak-Jaegermann, The cotype and uniform convexity of unitary ideals.

[10] R.K. Getoor and M.J. Sharpe, Conformal martingales. Invent. Math. 16 (1972)271-308.

[11] J. Globevnik, On complex strict and uniform convexity. Proc. Amer. Math. Soc. 47(1975)175-178.

[12] J. von Neumann, On rings of operators, reduction theory. Ann. of Math. (2) 50(1949)401-485.

[13] P. Noverraz, Pseudo-convexité, Convexité Polynomiale et Domaines d'Holomorphie en Dimension Infinie. Mathematics Studies No. 3. North-Holland 1973.

[14] J. Peetre, Locally analytically pseudo-convex topological vector spaces. Studia Math. 73(1982)253-262.

[15] T. Rado, Subharmonic functions. Springer-Verlag 1937.

[16] L. Schwartz, Semi-Martingales sur des Variétés, et Martingales Con-

formes sur des Variétés Analytiques Complexes. Lecture Notes in Math 780. Springer-Verlag 1980.

[17] E. Thorp and R. Whitley, The strong maximum modulus theorem for analytic functions into a Banach space. Proc. Amer. Math. Soc. 18 (1967)640-646.

LAWS OF RANDOM GAUSSIAN FUNCTIONS, SOME INEQUALITIES

X. Fernique
Département de Mathématique
Université Louis Pasteur
67084 Strasbourg Cedex
France

In 1975 C. Borell proved an important isoperimetric inequality concerning the laws of Gaussian vectors. Even if this inequality is apparently difficult to understand, it leads in fact to simple, manageable relations for the laws of Gaussian random functions. We show here some examples and their use.

0.0 In all that follows we use the language of random functions. We study random Gaussian functions (r.g.f.) defined on a metrizable, separable space T. We suppose that they are centered and separable. Let X be a r.g.f. on T and let d_X and σ_X be functions defined by:

$$\forall\ (s,t) \in T \times T\ ,\ d_X^2(s,t) = E|X(s) - X(t)|^2\ ,\ \sigma_X^2(t) = E|X(t)|^2\ ;$$

we will also use the following form of <u>metric entropy</u> : for every continuous pseudo-metric d on T and all $\delta > 0$, $S(T,d,\delta)$ is a set S of T, δ-distinguishable for d ($s \in S$, $t \in S$, $d(s,t) \le \delta \Rightarrow s = t$), with maximum cardinality $M = M(T,d,\delta)$ under this condition; the closed d-balls $B_d(s,\delta)$, $s \in S(T,d,\delta)$, form a covering for T. With this notation we recall:

PROPOSITION 0.1 *Let* X *be a r.g.f. a.s. bounded on* T. *Then*

0.1.1. $$\forall\ \delta > 0\ ,\ \delta\sqrt{\log M(T,d_X,\delta)} \le \delta + E[\sup_T X] < \infty\ ;$$

also, for every non-decreasing positive couple (δ,δ_o) :

0.1.2 $$\delta\sqrt{\log[M(T,d_X,\delta)]} \le \delta\sqrt{\log[M(T,d_X,\delta_o)]} + \sup_{s \in T} E[\sup_{d_X(s,t) \le \delta_o} X(t)] + \delta\ .$$

In particular, if X *is a.s. continuous for every point of* (T,d_X), *then*

0.1.3 $$\lim_{\delta \to 0} \delta\sqrt{\log M(T,d_X,\delta)} = 0 .$$

Elements of proof: The comparison properties for Gaussian vectors [2] show that $E \sup_T X$ is minorized by $\frac{\delta}{2} E[\sup^M \lambda_i]$ (where λ_i, $i \in [1,M(T,d_X,\delta)]$ is isonormal); simple calculations lead to 0.1.1 Property 0.1.2. results from the inequality

$$M(T,d_X,\delta) \le M(T,d_X,\delta_0) \sup_{s \in T} M(B_{d_X}(s,\delta_0),d_X,\delta)$$

and that of 0.1.1. applied to X on the balls $B_{d_X}(s,\delta_0)$, $s \in T$.

We now state a corollary of Borell's inequality suitably adapted to the study of r.g.f. :

PROPOSITION 0.2. Let X be a r.g.f. a.s. bounded on T and S a set of T. Then the median $m(\sup_S X)$ is unique and

0.2.0 $\forall m \ge m(\sup_S X)$, $\forall t \ge 0$, $P\{\sup_S X - m > t \sup_S \sigma_X\} \le \frac{1}{\sqrt{2\pi}} \int_t^\infty e^{-u^2/2} du$

0.2.1 $\forall t \ge 0$ $P\{|\sup_S X - m(\sup_S X)| > t \sup_S \sigma_X\} \le \sqrt{\frac{2}{\pi}} \int_t^\infty e^{-u^2/2} du$

and, in particular 0.2.2. $|m(\sup_S X) - E(\sup_S X)| \le \sqrt{\frac{2}{\pi}} \sup_S \sigma_X$.

Elements of proof: It suffices to prove 0.2.0 and 0.2.1. for every median of $\sup_S X$ where T is finite; the uniqueness of the median follows. Denote by A_1 the subset of $\mathbb{R}^T$ defined by $A_1 = \{x: \sup_S x \le m\}$ and in such a way that $P(A_1) \ge \frac{1}{2} = \frac{1}{\sqrt{2\pi}} \int_0^\infty e^{-u^2/2} du$. Note moreover that if h belongs to the unit ball 0_X of the autoreproducing space associated to X, then $\sup_S h \le \sup_S \sigma_X$ in such a way that

$$A_1 + t\, 0_X \subset \{x : \sup_S x \le m + t \sup_S \sigma_X\} .$$

The Borell inequality [1] applied to $(A_1,0)$ thus furnishes

$$P\{\sup_S X \le m + t \sup_S \sigma_X\} \ge \frac{1}{\sqrt{2\pi}} \int_{-\infty}^{t} e^{-u^2/2}\, du ,$$

and thus 0.2.0. in passing to the complement. We obtain 0.2.1. by using $A_2 = \{x : \sup_S x \ge m(\sup_S X)\}$ and similar calculations.

1. The following theorem provides the main inequalities of this paper; the underlined factors are the best possible.

THEOREM 1. _Let_ X _be r.g.f. on_ T _and_ d _a pseudo-metric on_ T _larger than or equal to_ d_X. _Then for every positive non-decreasing couple_ (δ_1,δ) _we have_

1.1 $$m\{\sup_{d(s,t)\le\delta} |X(s)-X(t)|\} \le \underline{\underline{2}} \sup_{s\in T} m\{\sup_{d(s,t)\le\delta_1} (X(s)-X(t)\} + $$
$$+ (\underline{\underline{\delta}}+4\delta_1)\sqrt{2\log M(T,d,\delta_1)\sup_{s\in T} M(B_d(s,\delta+2\delta_1),d,\delta_1)}$$

1.2 $$E\{\sup_{d(s,t)\le\delta} |X(s)-X(t)|\} \le \underline{\underline{2}} \sup_{s\in T} E\{\sup_{d(s,t)\le\delta_1} X(t)\} + \delta + 2\delta_1 +$$
$$+ (\underline{\underline{\delta}}+4\delta_1)\sqrt{2\log M(T,d,\delta_1)\sup_{s\in T} M(B_d(s,\delta+2\delta_1,d,\delta_1)}.$$

If, moreover, there exists a r.g.f. Y _associated with_ $d = d_Y$, _then for every positive non-decreasing couple_ (δ,δ_0):

1.3 $$E\{\sup_{d(s,t)\le\delta} |X(s)-X(t)|\} \le 13[\sup_{s\in T} E\{\sup_{d(s,t)\le\delta_0} Y(t)\} + \delta + \delta\sqrt{\log[M(T,d,\delta_0)]}]$$

Elements of proof: We use here the set $S = S(T,d,\delta_1)$ defined in 0.0. For every $t \in T$, there exists thus an element $s(t)$ of S such that $d(t,d(t)) \le \delta_1$. If $d(t,t') < \delta$ we have thus $d_X(s(t),s(t')) \le \delta+2\delta_1$. From this we deduce (abbreviating the notation of the second part of 1.1):

$$P\{\sup_{d(t,t')\le\delta} |X(t)-X(t') \ge 2m + (\delta+4\delta_1)\sqrt{2\log MM'}\} \le$$

$$\leq \sum_{s\in S} 2\,P\{\sup_{d(s,t)\leq\delta_1} (X(t)-X(s)) \geq m + \delta_1\sqrt{2\log MM'}\} +$$

$$+ \sum_{\substack{s\in S, s'\in S \\ d(s,s')\leq\delta+2\delta_1}} P\{X(s)-X(s') \geq d_X(s,s')\sqrt{2\log MM'}\} .$$

Using Borell's inequality 0.2.0, the laws of $X(s) - X(s')$, and computing the number of terms, we obtain (except for some small values of M and M')

$$P \leq 2M \frac{1}{\sqrt{2\pi}} \int_{\sqrt{2\log MM'}}^{\infty} e^{-u^2/2} du + MM' \frac{1}{\sqrt{2\pi}} \int_{\sqrt{2\log MM'}}^{\infty} e^{-u^2/2}\, du \leq \frac{1}{2} .$$

Direct calculations suffice for the other cases; this gives 1.1. We thus obtain 1.2. in using 0.2.2. To obtain 1.3. we recopy 1.2. using the crude form

$$E \sup_{d(s,t)\leq\delta} |X(s)-X(t)| \leq 2 \sup_{s\in T} E \sup_{d(s,t)\leq\delta} X(t) + 3\delta + 10\sqrt{\log M(T,d,\delta_1)} .$$

We majorize the first term of the second member by the term corresponding to Y, using the comparison properties. Applying 0.1.2. to Y we majorize the last term.

It remains to show that the underlined coefficients in 1.1. are the best possible. For this it suffices to use an isonormal sequence of length n with $\delta = \delta_1 = \sqrt{2}$, and then $\delta_1 < \delta = \sqrt{2}$. One could also use Brownian motion on $[0,1]$ with $\delta = h$, $\delta_1 = c\,h$, c small.

2. An application.

2.1. Let Y be a r.g.f. on T having continuous trajectories on (T,d_Y). Let $G_A(Y)$ be the set of Gaussian probabilities g on $C(T)$ such that

$$\text{(2.1.1)} \qquad \sup_{t\in T} \int |x^2(t)|\,dg(x) \leq A, \quad \int |x(s)-x(t)|^2\, dg(x) \leq d_Y^2(s,t) .$$

Property (1.3) shows that

$$\lim_{\delta\to 0} \sup_{g\in G_A(Y)} \int \sup_{d_Y(s,t)\leq\delta} |x(s)-x(t)|\,dg(x) = 0 .$$

G_A <u>is therefore a tight set</u>. As it is clearly closed for the weak topology, it is thus <u>weakly compact</u>. This powerful property is well suited for the study of limit theorems (the central limit theorem, for example [3]).

2.2 For the last 10 years it has been known [2] that the relation 2.1.1. implies

$$\forall\ S \subset T \qquad \int \sup_{t \in S} x(t)\, dg(x) \le E \sup_{t \in S} Y(t) ,$$

and thus the set G_A is in a certain sense a bounded set; this property has already been used to study the geometry of certain Banach spaces. The compactness property presented here could well have more extensive applications.

<u>REFERENCES</u>

[1] C. Borell, Inv. Math., 30, 1974, 207-211.

[2] X. Fernique, Springer Lecture Notes 480, 1975, 1-96.

[3] E. Gine, and J. Zinn, Some Limit Theorems for Empirical Processes, Ann. Prob., Ann. Prob. 1984, Vol 12, No 4, 929-989.

ON THE RADON-NIKODYM PROPERTY IN FUNCTION SPACES

N. Ghoussoub
Department of Mathematics
University of British Columbia
Vancouver, British Columbia
CANADA

and

B. Maurey
Universite´ Paris VII
Paris, France

Abstract: Using the topological methods introduced in [3], we give a simple proof of a theorem of Talagrand [7] which asserts that Banach lattices with the Radon-Nikodym property are isometric to dual Banach lattices. We also give a proof of a recent result announced by Diestel [2]: namely that the point of continuity and the Radon-Nikodym properties are equivalent for closed convex bounded subsets of Banach lattices not containing c_o. The case of subsets of L_1 was first observed by Bourgain-Rosenthal [1].

Introduction: Let X be a Banach space. Recall that a subset C of X is said to have the point of continuity property (resp. the Radon-Nikodym property) if every weakly closed bounded subset of C has a point of weak to norm continuity (resp. a denting point).

The authors proved in [3], that separable Banach spaces with the above defined properties "live" in dual spaces as w^*-G_δ subspaces. On the other hand, if X is a Banach lattice not containing c_o, then X can be represented as a function space on some probability space (Ω,F,P) in such a way that X is an order ideal (non-closed) in $L_1(\Omega,F,P)$. In order to prove the assertions mentioned in the abstract we start by combining these two representations in a fashion suitable for our study.

Proposition (1): Let X be a separable Banach lattice not containing c_o and let C be a closed bounded convex subset of X with the point of continuity property; then there exist a separable Banach lattice Y, a metrizable compact Hausdorff space Ω, a Radon probability measure μ on Ω and a dense range lattice homomorphism $S:C(\Omega) \to Y$ such that:

(i) X is lattice isometric to a closed order ideal of Y^* in such a way that C is a w^*-G_δ in Y^*.

(ii) The restriction of the operator $S^*:Y^* \to M(\Omega)$ to X maps it into $L_1(\Omega,\mu)$ is such a way that

(a) $S^{*-1}(L_1(\Omega,\mu) \subseteq X$

(b) $L_\infty(\Omega,\mu)$ is an order ideal of $S^*(X)$.

Moreover, if C has the Radon-Nikodym property then Y can be chosen in such a way that $Y^*\backslash C = \bigcup_n K_n$ where each K_n is w*-compact and convex.

Proof: By Lemma II.2 of [3], C can be written as $\bigcap_n (K_n \cup O_n)$ where each K_n is w*-compact in X^{**} and each O_n is a countable union of w*-open elementary neighborhoods of X^{**}. Let Y be the smallest separable sublattice of X^* containing a norming subset for X and all the functionals determining the w*-open sets (O_n). Let T be the canonical injection of Y into X^*. We get as in Lemma II.3 of [3] that the restriction of the quotient map $T^*: X^{**} \to Y^*$ to the space X is an isometry that maps the set C into a w*-G_δ of Y^*. On the other hand, T is a lattice isometry, hence T^* is interval preserving [6]; since X is an order ideal in X^{**} [5], we get that $T^*(X)$ is an order ideal in Y^*. We shall now identify X with $T^*(X)$ which clearly verifies the assertion (i) of the proposition.

Let now u_o be a norm one quasi-interior point of Y and let $C(\Omega_o)$ be the A-M space obtained by norming the vector lattice $\bigcup_n [-nu_o, nu_o]$ with the gauge of the order interval $[-u_o, u_o]$. Note that the canonical injection $S: C(\Omega_o) \to Y$ is a dense range, lattice homomorphism which is also interval preserving. Let $C(\Omega)$ be the separable sublattice of $C(\Omega_o)$ generated by a countable dense subset of $[-u_o, u_o]$ in view of the separability of Y.

Note now that $S^*: Y^* \to M(\Omega)$ is an interval preserving lattice homomorphism which is also one-to-one. Since X is separable, consider a quasi-interior point x_o of X such that $u_o(x_o) = 1$ and set $\mu = T^*x_o$. It is clear that $L_\infty(\Omega,\mu) \subseteq S^*(X) \subseteq L_1(\Omega,\mu)$.

To prove (ii)(a) note that if $y^* \in Y^*_+$ is such that $S^*y^* \in L_1(\Omega,\mu)$ then $S^*y^* \wedge n = S^*(y^* \wedge nx_o)$ converges in L_1 to S^*y^*, hence $S^*(y^* \wedge nx_o) \xrightarrow{w^*} S^*y^*$ and $y^* \wedge nx_o$ w*-converges to y^* since S^* is a w*-homeomorphism on the bounded subsets of Y^*. But the sequence $(y^* \wedge nx_o)$ is increasing and norm-bounded in X hence it converges in norm to an element in X which must be y^*.

We can now prove the following theorem due to Talagrand [7].

Theorem (1): A separable Banach lattice has the Radon-Nikodym property if and only if it is isometric to the dual of separable Banach lattice.

Proof: Apply Proposition (1) to the ball of X to get Y such that X is isometric to an order ideal of Y^* in such a way that $Y^*\backslash B_X = \underset{n}{\cup} K_n$ where each K_n is w*-compact and convex. Let $S: C(\Omega) \to Y$ be as in the conclusion of the proposition: that is $L_\infty(\Omega,\mu) \subseteq S^*X \subseteq L_1(\Omega,\mu)$ for some probability measure μ on Ω.

We shall prove that for all $\varepsilon > 0$, there exists a measurable subset $\Omega_\varepsilon \subseteq \Omega$, with $\mu(\Omega_\varepsilon) \geq 1-\varepsilon$, and such that the set $B_\varepsilon = \{x \in B_X;\ S^*x = 0$ on $\Omega\backslash\Omega_\varepsilon\}$ is w*-compact in B_{Y^*}. This would prove the claim since, as noted by Talagrand [7] one then gets by exhaustion that X is a boundedly complete disjoint sum of dual Banach lattices.

Fix now $\varepsilon > 0$ and note that for each n, we have $(\frac{2^n}{\varepsilon})B_{L_\infty} \subseteq S^*(X)$ hence it is w*-compact and disjoint from $S^*(K_n)$. There exists then $\phi_n \in C(\Omega)$ with $\phi \leq 1$ on $(\frac{2^n}{\varepsilon})B_{L_\infty}$ and $\phi_n > 1$ on K_n.

It follows that $\|\phi_n\|_1 \leq \varepsilon\, 2^{-n}$ and if we set $\Omega_n = \{w \in \Omega : |\phi_n| \leq 1\}$ we get $\mu(\Omega_n) \leq \varepsilon\, 2^{-n}$.

Let now $\{x_k\}$ in B_X such that $S^*x_k = 0$ on $\Omega\backslash\Omega_n$. Suppose $x_k \xrightarrow{x^*} y^*$ for some $y^* \in Y^*$. We have

$$|\langle S^*x_k, \phi_n\rangle| = \left|\int_{\Omega_n} S^*x_k\phi_n d\mu\right| \leq \|S^*x_k\|_1 \leq \|x_k\| \leq 1$$

hence $\langle S^*y^*, \phi_n\rangle \leq 1$ and $y^* \notin K_n$.

Let now $\Omega_\varepsilon = \bigcap_{n\geq 1} \Omega_n$. We get in the same way that if $x_k \in B_X$, $S^*x_k = 0$ on $\Omega\setminus\Omega_\varepsilon$ and $x_k \xrightarrow{w^*} y^*$ then $y^* \notin \bigcup_n K_n$ hence $y^* \in B_X$. The claim is now proved since $\mu(\Omega_\varepsilon) \geq 1-\varepsilon$.

Now, we deal with the equivalence of the P.C and R.N properties for subsets of Banach lattices not containing c_o. These results were announced by Diestel [2]. We first start with the case of L_1 which was noticed first by Bourgain-Rosenthal [1].

For a bounded subset A of L_1, define for each $\varepsilon > 0$, $\delta_\varepsilon(A) = \sup\{\int_E |f|d\mu;\ f \in A \text{ and } \mu(E) \leq \varepsilon\}$. The modulus of equi-integrability of A is then:

$$\delta(A) = \lim_{\varepsilon\to 0} \delta_\varepsilon$$

The following lemma summarizes the properties of δ. The proof is left to the interested reader.

<u>Lemma (1)</u>: Let A be a bounded subset of $L_1(\Omega,\mu)$; then:

(i) $\delta(A + g) = \delta(A)$ for each g in $L_1(\Omega,\mu)$

(ii) $2\delta(A) \leq \operatorname{diam}(A)$.

(iii) For each λ in the w*-closure $\bar{A}^*$ of A in $M(\Omega)$, we have $d(\lambda, L^1) \leq \delta(A)$.

(iv) For each family $(A_i)_{i=1}$ of bounded subsets of L_1 and each sequence $(\theta_i)_{i=1}^n$, $\theta_i \geq 0$, with $\sum_{i=1}^n \theta_i = 1$ we have

$$\delta\left(\sum_{i=1}^n \theta_i A_i\right) \geq \sum_{i=1}^n \theta_i \delta(A_i).$$

<u>Remark</u>: The proof of (iii) is an immediate application of the subsequence splitting lemma.

<u>Lemma (2)</u>: Let D be a convex bounded subset of L_1 which is a w*-G_δ

subset of $M(\Omega)$; then for each $\varepsilon > 0$, there exists a w*-open slice S of D such that for each $\lambda \in \overline{S}^*$, $d(\lambda, L_1) \le \varepsilon$.

Proof: Since D is norm separable and a w*-G_δ subset of $M(\Omega)$, a classical theorem of Baire insures the existence of a w*-open set U in $M(\Omega)$ such that $D \cap U \neq \phi$ and $\operatorname{diam}(D \cap U) \le \varepsilon$. By Lemma 4 of [1], there exist w*-open slices $S_1, \ldots, S_n$ of D such that

$$\sum_{i=1}^{n} \theta_i S_i \subseteq D \cap U \quad \text{where} \quad \theta_i \ge 0 \ \forall\, i \quad \text{and} \quad \sum_{i=1}^{n} \theta_i = 1.$$

Note now that

$$\sum_{i=1}^{n} \theta_i \delta(S_i) \le \delta\Big(\sum_{i=1}^{n} \theta_i S_i\Big) \le \delta(D \cap U) \le \operatorname{diam}(D \cap U) \le \varepsilon.$$

Hence there exists S_{i_o} such that $\delta(S_{i_o}) \le \varepsilon$. It follows that for each $\lambda \in \overline{S}^*_{i_o}$, $d(\lambda, L_1) \le \varepsilon$.

Lemma (3): Let C_1 be a convex bounded subset of $L_1(\Omega,\mu)$ which is also a w*-G_δ in $M(\Omega)$. Then, there exists a sequence (K_n) of w*-compact convex subsets of $M(\Omega)$ and a sequence of w*-open subsets (O_n) consist- of countable unions of w*-open half-spaces of $M(\Omega)$ such that:

$$C_1 \subseteq \bigcap_n (K_n \cup O_n) \subseteq L_1(\Omega,\mu) \cap \overline{C}_1^*.$$

Proof: Fix $\varepsilon > 0$ and define inductively a decreasing family of relatively w*-closed subsets of C_1, which are w*-G_δ in $M(\Omega)$, in the following way:

(i) $F_o = C_1$

(ii) If $\alpha = \beta + 1$ and $F_\beta \neq \phi$ use Lemma (2) to find a w*-open slice S_β of F_β such that for each $\lambda \in \overline{S}^*_\beta$, $d(\lambda, L_1) \le \varepsilon$. Set $F_\alpha = F_\beta \backslash S_\beta$

(iii) If α is a limit ordinal, let $F_\alpha = \bigcap_{\beta<\alpha} F_\beta$.

Since the ball of $M(\Omega)$ is w*-metrizable, there exists $\gamma_\varepsilon < \Omega$ (the first uncountable ordinal) such that $F_{\gamma_\varepsilon} = \phi$. Let K_α be the w*-closure of F_α in $M(\Omega)$ and let H_α be the w*-open half-space such that $S_\alpha = H_\alpha \cap F_\alpha$. It is clear that

$$C_1 \subseteq \bigcap_{\alpha \le \gamma_\varepsilon} (K_\alpha \cup \bigcup_{\beta<\alpha} H_\beta).$$

Moreover, if x belongs to the right-hand side, then $x \in K_\beta \cap H_\beta$ for some $\beta \le \gamma_\varepsilon$, hence $x \in \overline{S}_\beta^{*}$ and $d(x,L_1) \le \varepsilon$. It follows that if we repeat the construction for each $\varepsilon = \frac{1}{n}$ we would get:

$$C_1 \subseteq \bigcap_n \bigcap_{\alpha \le \gamma_n} (K_{\alpha,n} \cup \bigcup_{\beta<\alpha} H_{\beta,n}) \subseteq \overline{C}_1^{*} \cap L_1(\Omega,\mu).$$

Now, we can prove the following.

<u>Theorem (2)</u>: Let X be a Banach lattice not containing c_o. Then, a separable closed convex bounded subset C of X has the Radon-Nikodym property if and only if it has the point of continuity property.

<u>Proof</u>: Suppose C has the P.C property. Apply Proposition (1) to find Y such that C is a w*-G_δ in Y^*. Note that $C_1 = S^*(C)$ is a w*-G_δ in $M(\Omega)$ which is contained in $L_1(\Omega,\mu)$. Apply Lemma (3) to C_1 to get that

$$C_1 \subseteq \bigcap_n (K_n \cup O_n) \subseteq \overline{C}_1^{*} \cap L_1(\Omega,\mu)$$

where each K_n is w*-compact and convex in $\overline{C}_1^{*}$ and each O_n is a countable union of w*-open half-spaces in $M(\Omega)$. Since S^* is one-to-one and $S^{*-1}(L_1(\Omega,\mu)) \subseteq X$, we get that

$$C_1 \subseteq D = \bigcap_n (S^{*-1}(K_n) \cup S^{*-1}(O_n)) \subseteq X .$$

But $S^{*-1}(K_n)$ is w*-compact and convex in Y^*, hence we can write $S^{*-1}(K_n) = \bigcap_m L_{n,m}$ where each $L_{n,m}$ is a w*-open half-space in Y^*. It follows that $Y^*\backslash D = \bigcup_n H_n$ where each H_n is w*-compact and convex in Y^*.

Hence D is w^*-H_δ-set [3,4] and every C-valued martingale w^*-converges a.e. to a w^*-measurable random variable valued a.e. in D. Since $D \subseteq X$, it is norm separable, hence the martingale converges strongly and is therefore valued in C which then has the Radon-Nikodym property.

Corollary (1): A separable Banach lattice has the point of continuity property if and only if it is isometric to the dual of a Banach lattice.

Proof: Note that in this case, the lattice does not contain c_o, hence Theorem (2) and then Theorem (1) apply to get the result.

REFERENCES

[1] J. Bourgain, H.P. Rosenthal: "Geometrical implications of certain finite dimensional decompositions". Bull. Soc. Math. Belg. 32 p.57-82 (1980).

[2] J. Diestel: Personal communication (1983).

[3] N. Ghoussoub, B. Maurey: "G_δ-embeddings in Hilbert space", Journal of Funct. Analysis Vol 61, No. 1, p. 72-97 (1985)

[4] N. Ghoussoub, B. Maurey: "H_δ-embeddings in Hilbert space and optimization on G_δ-sets", to appear in Memoirs of the A.M.S (1985)

[5] Y. Lindenstrauss, L. Tzafriri: "Classical Banach spaces. II. Function spaces, Springer-Verlag 97 (1979).

[6] H.P. Lotz: "Extensions and liftings of positive linear operators", T.A.M.S. 211, p. 85-100 (1975).

[7] M. Talagrand: "La structure des spaces de Banach reticules ayant la propriete de Radon-Nikodym", Israel J. Math. 44 No. 3, (1983).

SYMMETRIC SEQUENCES IN L^p, $1 \le p < +\infty$.

Sylvie Guerre
Equipe d' Analyse
Université Paris 6
4,Place Jussieu
75230 Paris Cedex 05

In their book: "Symmetric structures in Banach spaces" [2], W.B. Johnson, B. Maurey, G. Schechtman and L. Tzafriri asked the following question: "Does there exist a constant K such that for all p, $2 \le p < +\infty$, every normalized and weakly null sequence in L^p, has a subsequence which is K-symmetric?" They gave a partial answer: Fix $\varepsilon > 0$, then every normalized and weakly null sequence in L^p for $p \in 2\mathbb{N}$ has a $2(1+\varepsilon)$-symmetric subsequence. On the other hand, H.P. Rosenthal proved that, if $\varepsilon > 0$ is given, then every normalized sequence in L^1, which is equivalent to the unit vector basis of l^2, has a $2(1+\varepsilon)$-symmetric subsequence.

Here, as L^p-spaces are stable, we use the techniques of stability, introduced by J.L. Krivine and B. Maurey in [4], to generalize and unify these results.

THEOREM

Fix $\varepsilon > 0$. Then every normalized sequence in L^p which is weakly null if $p \ge 2$ and equivalent to the unit vector basis of l^2 if $1 \le p < 2$, has a subsequence which is $2(1+\varepsilon)$-symmetric.

REMARK

If $p > 2$, we know by Kadec and Pelczynski's theorem [3], that if (x_n) is a normalized and weakly null sequence in L^p,

- either (x_n) has a subsequence which is $(1+\varepsilon)$-equivalent to the unit vector basis of l^p
- or (x_n) is equivalent to the unit vector basis of l^2 .

In the first case, the theorem is obviously true. So, this theorem gives a property of l^2-sequences in L^p.

SKETCH OF PROOF

Let (x_n) be a normalized sequence in L^p, $1 \le p < +\infty$, which is equivalent to the unit vector basis of l^2. Let σ be a type defined by (x_n) (i.e., $\forall x \in L^p$, $\sigma(x) = \lim \|x + x_n\|$). Then, we know (J.L. Krivine and

B. Maurey) that if the conic class $K_1(\sigma)$ generated by σ(i.e., $K_1(\sigma) = \{$types Z on L^p, such that $Z = a_1 \sigma * \ldots * a_k \sigma$, $k \in \mathbb{N}$, $(a_1,\ldots,a_k) \in \mathbb{R}^k$, $\|Z\| \le 1\}$) is compact for the topology of uniform convergence on bounded sets in L^p, then (x_n) has a subsequence which is $(1+\varepsilon)$-equivalent to the fundamental sequence of the spreading model associated to (x_n) or σ(i.e., $\mathbb{R}^{\mathbb{N}}$ equipped with the norm : $\|(a_1,\ldots,a_k)\| = \|a_1\sigma * \ldots * a_k\sigma\|$). See [1] for a more precise definition]. As any spreading model on a stable space is 2-symmetric, this subsequence is $2(1+\varepsilon)$-symmetric and the conclusion of the theorem holds.

We use a representation of the functions $\|x+y\|^p$ in L^p-spaces that gives us a description of all types on L^p, $1 \le p < +\infty$, and ensures that, under the hypothesis of the theorem, $K_1(\sigma)$ has the suitable property.

OPEN PROBLEMS

- What can we say about sequences in L^p, $1 \le p < 2$, which are not equivalent to the unit vector basis of l^2?
- Does there exist an equivalent norm on l^2, and a normalized and weakly null sequence in l^2, which does not have a $2(1+\varepsilon)$-symmetric subsequence for some $\varepsilon > 0$ in that norm?

REFERENCES

[1] A. Brunel and L. Sucheston, On B-convex Banach spaces, Math. Syst. Theory, t. 7, no 4, 1973.

[2] W.B. Johnson, B. Maurey, G. Schechtman, L. Tzafriri, Symmetric structures in Banach spaces, Memoirs of the A.M.S., May 1979, Vol. 19, no. 217.

[3] M.I. Kadec and A. Pelczynski, Bases, lacunary sequences and complemented subspaces in L^p Studia Math. t. 21 (1962).

[4] J.L. Krivine and B. Maurey, Espaces de Banach stables, Israel J. of Math., Vol. 39, no. 4, 1981.

ON THE LOCAL STRUCTURE OF $L_p(X)$

Richard Haydon, Mireille Levy and Yves Raynaud
Department of Mathematics
Brasenose College
Oxford, England OX1 4AJ

1. INTRODUCTION

In this note, we consider a question mentioned to us originally by Gilles Pisier: If X,Y are Banach spaces and Y is finitely representable in $L_p(X)$, then is Y necessarily embeddable in $L_p(\mu;\tilde{X})$ for some measure μ and some Banach space $\tilde{X}$ which is itself finitely representable in X ? An affirmative answer in the special case where X is of the form $L_q(\nu)$ was given in [4]. (For separable Y, this result was known to Heinrich [3] and to Krivine.) Here we sketch a proof that the answer in general is in the negative. The results will be presented in greater detail as a part of a longer work on randomly normed spaces [1].

Our notation and terminology are standard. We recall that a Banach space Y is said to be finitely representable in Z if, for every $\varepsilon > 0$ and every finite-dimensional subspace F of Y, there is a subspace G of Z with $d(F,G) < 1 + \varepsilon$. The important relationship between finite representability and ultraproducts is that Y is f.r. in Z if and only if Y embeds isometrically in some ultrapower of Z [2].

2. RANDOM BANACH SPACES AND DIRECT INTEGRALS OF L_q-SPACES

We start by introducing a definition which gives us a useful framework in which to study ultraproducts of $L_p(X)$ and similar spaces.

Definition. Let (Ω,Σ,μ) be a measure space and Z be a module over $L_\infty(\Omega,\Sigma,\mu)$. A random norm on Z, with values in $L_p(\mu)$, is a map $N: Z \to L_p^+(\mu)$ which satisfies:

(i) $N(w+z) \le N(w) + N(z)$

(ii) $N(\phi\cdot z) = |\phi|\cdot N(z) \qquad (\phi \in L_\infty(\mu))$

(iii) $N(z) = 0$ if and only if $z = 0$.

If a randomly normed space Z is complete for the associated (scalar) norm, $\|z\| = \|N(z)\|_{L_p(\mu)}$, we say that Z is a random Banach space over $L_p(\mu)$.

The most obvious example of a random Banach space over $L_p(\mu)$ is the space $L_p(\mu;X)$ of all (Bochner) μ-integrable functions with values in the Banach space X. The random norm is, of course, given by $N(f)(\omega) = \|f(\omega)\|$. In the case where X is of the form $L_p(\Omega',\Sigma',\mu')$ and $1\le p,q<\infty$, $L_p(\mu;X)$ can be regarded as the space of all (equivalence classes of) measurable functions f on $\Omega \times \Omega'$ for which the following norm is finite

$$\|f\| = \left(\int_{\Omega}\left(\int_{\Omega'} |f(\omega,\omega')|^q d\mu'(\omega')\right)^{p/q} d\mu(\omega)\right)^{1/p}.$$

This observation leads us to introduce a more general class of random Banach space.

<u>Definition</u>. Let (Ω,Σ,μ), (Ω',Σ',μ') be measure spaces, Let $p \ge 1$ be a real number and let $q:\Omega\to[1,\infty)$ be a Σ-measurable function. We define the <u>direct integral</u>

$$\left(\int_{\Omega}^{\otimes} L_{q(\omega)}(\mu')\, d\mu(\omega)\right)_p$$

to be the space of all measurable functions f on $\Omega \times \Omega'$ for which

$$\|f\| = \left(\int_{\Omega}\left(\int_{\Omega'} |f(\omega,\omega')|^{q(\omega)} d\mu'(\omega')\right)^{p/q(\omega)} d\mu(\omega)\right)^{1/p}$$

is finite. This a random Banach space over $L_p(\mu)$ when equipped with the random norm

$$N(f)(\omega) = \|f(\omega,\cdot)\|_{L_{q(\omega)}(\mu')}.$$

The following result, which extends a theorem given in [4], shows that an intrinsic property of the random norm 'almost' characterizes those random Banach <u>lattices</u> which are direct integrals of L_q-spaces.

<u>Theorem</u> <u>2.1</u>. Let (Ω,Σ,μ) be a measure space, $p \ge 1$ be a real number and $q:\Omega\to[1,\infty)$ be Σ-measurable. Let Z be a random Banach lattice over $L_p(\mu)$ having the property that, whenever w,z are disjoint elements of Z, we have

$$N(w+z)(\omega) = \left(N(w)(\omega)^{q(\omega)} + N(z)(\omega)^{q(\omega)}\right)^{1/q(\omega)}$$

for almost all ω in Ω . Then there exists a measure space (Ω',Σ',μ') such that Z is isometrically lattice isomorphic to a band in the direct integral

$$\left(\int_{\Omega}^{\otimes} L_{q(\omega)}(\mu')\, d\mu(\omega)\right)_p$$

<u>Remarks</u>. (i) If in the above theorem one wishes to be able to conclude that Z is identifiable with the whole of some direct integral, rather than just a band, then some additional hypothesis (of 'homogeneity') is needed.

(ii) An alternative way to clean up the conclusion of 2.1 is to extend the definition of direct integral to allow the possibility of varying measures μ'_ω .

3. ULTRAPRODUCTS OF RANDOM BANACH SPACES

It is well known that an ultraproduct of a family of spaces $L_p(\mu_i)$ is identifiable with $L_p(\mu)$ for some measure μ , provided $p < \infty$. This result leads to a stability result for random Banach spaces over L_p.

<u>Theorem 3.1</u>. Let I be a set, U be an ultrafilter on I and, for each $i\in I$, let $(\Omega_i,\Sigma_i,\mu_i)$ be a measure space. Let $p \geq 1$ be a fixed real number and, for each i let Z_i be a random Banach space over $L_p(\mu_i)$. If Z is the (Banach space) ultraproduct $(\Pi Z_i)/_U$ and $L_p(\mu)$ is the ultraproduct $(\Pi L_p(\mu_i))/_U$, then Z can be equipped with the structure of a random Banach space over $L_p(\mu)$.

<u>Sketch of proof</u>. If N_i is the random norm on Z_i and $(z_i)^\bullet$ denotes the element of Z corresponding to the family $(z_i)_{i\in I}$, we may define $N:Z\to L_p^+(\mu)$ by writing $N((z_i)^\bullet) = (N_i(z_i))^\bullet$. It is not immediately clear that Z has the structure of an $L_\infty(\mu)$-module, though for $\phi = (\phi_i)^\bullet \in (\Pi L_\infty(\mu_i))/_U$, we may define $\phi.(z_i)^\bullet = (\phi_i.z_i)^\bullet$. The ultraproduct $(\Pi L_\infty(\mu_i))/_U$ is a weak*-dense subalgebra of $L_\infty(\mu)$ and an approximation argument enables us to extend the module structure to the whole of $L_\infty(\mu)$.

<u>Theorem 3.2</u>. In the situation considered in 2.1 assume further that, for

each i, Z_i is a direct integral $(\int_{\Omega_i}^{\otimes} L_{q_i(\omega)}(\mu_i') \, d\mu_i(\omega))_p$, where the functions q_i are such that $\sup_i \sup_\omega q_i(\omega) < \infty$. Then the ultraproduct Z is isometrically lattice-isomorphic to a band in some direct integral $(\int_\Omega^{\otimes} L_{q(\omega)}(\mu') \, d\mu(\omega))_p$.

Moreover, if $\alpha \leq q_i(\omega) \leq \beta$ for all i and all $\omega \in \Omega_i$, then $\alpha \leq q(\omega) \leq \beta$ for all $\omega \in \Omega$.

<u>Sketch of proof</u>. We consider the element $q = (q_i)^{\cdot}$ of $(\Pi L_\infty(\mu_i))/U$ which, as before, we regard as being contained in $L_\infty(\mu)$. What has to be shown is that the hypotheses of 2.1 are now satisfied.

4. THE EXAMPLE

In this section we sketch a proof that, for any $p > 2$, there are Banach spaces X and Y such that Y is finitely representable in $L_p(X)$ while Y does not embed isometrically in any space $L_p(\tilde{X})$ with $\tilde{X}$ finitely representable in X. In fact, for simplicity of notation, we choose to work with the specific case $p = 4$, but the general case proceeds in exactly the same manner.

We take Y to the direct integral $(\int_{[2,3]}^{\otimes} L_q[0,1] dq)_4$ so that Y contains Lebesgue measurable functions f on $[2,3] \times [0,1]$ for which $\|f\| = (\int_2^3 (\int_0^1 |f(q,t)|^q dt)^{4/q} dq)^{4/q}$ is finite. It is not hard to see that Y is finitely representable in $L_4(X)$ whenever X is a space such that ℓ_q is finitely representable in X for all $2 \leq q \leq 3$. For our purposes, we may take X to be either the direct integral $(\int_{[2,3]}^{\otimes} L_q[0,1] dq)_3$ or the direct sum $(\bigoplus_{2 \leq q \leq 3, q \in \Omega} \ell_p)_3$. In either case we deduce from 3.2 that any space X which is finitely representable in X embeds isometrically in some direct integral of the form $(\int_\Omega^{\otimes} L_{q(\omega)}(\mu') \, d\mu(\omega))_3$ where $2 \leq \omega \leq 3$ for all $\omega \in \Omega$.

Assume, if possible, that Y embeds isometrically in $L_4(\tilde{X})$ for some such $\tilde{X}$. Since Y is separable and each member of $L_4(\tilde{X})$ almost surely takes values in a separable subspace of X, we may suppose that X is separable. Hence, our assumption implies that there exists an isometric embedding $T: Y \to L_4(\Omega,\Sigma,\mu;Z)$ where Z is a direct integral $(\int_\Omega^\otimes L_{q(\omega')}(\Omega'',\Sigma'',\mu'')\,d\mu'(\omega'))_3$, $2 \leq q(\omega) \leq 3$, and all of (Ω,Σ,μ), (Ω',Σ',μ') , $(\Omega'',\Sigma'',\mu'')$ are separable probability spaces.

Making use of the assumption that T is an isometry, we can show that if y_1,y_2 are elements of Y which are supported by disjoint subsets of $[2,3]$ then the elements Ty_1,Ty_2 are supported by disjoint subsets of Ω (since in this case $\|Ty_1 \pm Ty_2\|^4 = \|Ty_1\|^4 + \|Ty_2\|^4$). This is the first step in showing that we may assume Ω to be of the form $[2,3]\times\overline{\Omega}$, μ to be the product of Lebesgue measure on $[2,3]$ with some measure $\overline{\mu}$ to $\overline{\Omega}$, and T to be 'modular' in the sense that $T(\phi\cdot y) = \phi\cdot Ty$ for $y \in Y$ and $\phi \in L_\infty[2,3]$.

Having shown that T is a isometry from Y into $L_4([2,3];L_4(\mu;Z))$, one next shows that for almost all $q \in [2,3]$ there is an isometry $T_q: L_q[0,1] \to L_4(\overline{\mu};Z)$ such that T is derived from the familty (T_q) via the formula $Ty(q) = T_q(y(p,\cdot))$ (for all $y \in Y$ and almost all $q \in [2,3]$. Hence, we have reached the situation where $L_4(\overline{\mu};Z)$ has, for almost all $p \in [2,3]$, a subspace isometric to L_p.

On the other hand, using arguments of Kadec-Pelczynski type and some simple cotype estimates, we are able to show that if $L_4(\overline{\mu};Z)$ has a subspace isomorphic to ℓ_p and $2 < p < 3$ then we must have

$$\mu'\{\omega' \in \Omega : q(\omega') = p\} > 0.$$

Since we had already reduced ourselves to the case where μ' is a finite measure, this is a contradiction.

<u>Remark</u>. The final stage in the argument sketched above can be deduced from Heinrich's result about the Isomorphic Envelope Problem [3]. Heinrich shows (without obtaining a concrete representation for such spaces) that if a separable space X is finitely representable in the direct sum $(\bigoplus_{2 \leq q \leq 3, q \in \mathbb{Q}} \ell_q)_4$ then X can have subspaces isomorphic to

ℓ_p for only countably many values of p.

Open Problems. (i) We do not know whether, in the example above, there exists an isomorphic embedding of Y in $L_4(\tilde{X})$ for some $\tilde{X}$ f.r. in X.

(ii) The question posed at the beginning of this paper remains open for values of p in the range $1 \leq p \leq 2$. We suspect that the answer is again in the negative, but the construction used here certainly does not provide a counterexample.

REFERENCES

[1] R.Haydon, M. Levy and Y. Raynaud, Randomly normed spaces, (in preparation).

[2] S. Heinrich, Ultraproducts in Banach space theory, J. Reine Angew. Math. 313 (1980), 72-104.

[3] S. Heinrich, The isomorphic problem of envelopes, Studia Math. 73 (1982), 41-49.

[4] M. Levy and Y. Raynaud, Ultrapuissances de $L_p(L_p)$', Comptes Rendus Acad. Sci. Paris 299 (1984), Serie I, 81-84.

ON WEAKLY COMPACT OPERATORS ON $\mathscr{C}(K)$-SPACES

Hans Jarchow and Urs Matter
Institut für Angewandte Mathematatik
Universität Zürich, Rämistrasse 74
CH 8001 Zürich, Switzerland

1. Introduction

Several operator ideals are known which have the property that their components consist precisely of the weakly compact operators, provided that the domain space is chosen as a $\mathscr{C}(K)$-space, the Banach space of all continuous functions on a compact Hausdorff space K. We only mention the ideals of fully complete operators, of unconditionially summing operators, of strictly singular operators, etc. We refer for this to [12], [13].

In contrast to the ideals just mentioned which are, in a sense, fairly large, we intend to discuss in this note much smaller ideals with the above property, and some of their relatives. They are all connected with the so-called absolutely continuous operators introduced by C.Niculescu [11]. Among other things, we shall see that these ideals are stable with respect to the formation of ultrapowers, a fact which allows, for example, easy proofs of some well-known results on super-reflexivity.

Some of the material presented here will be part of the dissertation [8] of the second named author written under the direction of the first at the University of Zürich.

2. Notation

We shall employ standard Banach space notation. X,Y,Z,... will be used to denote Banach spaces. If X is a Banach space, then I_X, B_X, and X* will be, respectively, its identity map, its closed unit ball, and its continuous dual. Sometimes, we will consider B_{X^*} as a compact space with respect to the weak* topology of X*. By a subspace of a Banach space, we always mean a closed linear submanifold, and continuous linear maps between Banach spaces are simply referred to as operators. The reader may consult [7] for further unexplained notations and facts.

As for ideals of operators, our standard reference is of course [16]. We shall freely make use of the notations and results given there. The most important ideals for us will be those of compact, weakly compact, fully complete, absolutely p-summing, p-integral, and p-factorable operators, which we denote by $\mathcal{K}$, $\mathcal{W}$, $\mathcal{V}$, $\mathcal{P}_p$, $\mathcal{I}_p$, and Γ_p, respectively. If $\mathcal{A}$ and $\mathcal{B}$ are ideals, then we write $\mathcal{A}(X,\cdot)\subset\mathcal{B}(X,\cdot)$ if $\mathcal{A}(X,Y)$ is contained in $\mathcal{B}(X,Y)$ for all Banach spaces Y, etc. Further, $\mathcal{A}\circ\mathcal{B}$ is the ideal formed by all operators $A\circ B$ with A and B belonging to (appropriate components of) $\mathcal{A}$ and $\mathcal{B}$, respectively. In place of $\mathcal{A}\circ\mathcal{A}$, we prefer to write $\mathcal{A}^2$.

3. Absolutely continuous operators

Let $\mathcal{A}$ be any operator ideal. Its uniform closure, $\bar{\mathcal{A}}$, is the ideal which is obtained by taking for $\bar{\mathcal{A}}(X,Y)$ just the closure of $\mathcal{A}(X,Y)$ in $\mathcal{L}(X,Y)$ with respect to the uniform norm topology. $\mathcal{A}$ is uniformly closed if $\mathcal{A}=\bar{\mathcal{A}}$.

The injective hull of an ideal $\mathcal{A}$ is denoted $\mathcal{A}^{inj}$. It is easy to see that $\bar{\mathcal{A}}^{inj}$ (the injective hull of $\bar{\mathcal{A}}$) is uniformly closed. The following characterization of $\bar{\mathcal{A}}^{inj}$ was proved in [5]:

(1) An operator $T:X\longrightarrow Y$ belongs to $\bar{\mathcal{A}}^{inj}(X,Y)$ if and only if, for every $\varepsilon>0$, there exists a Banach space Z_ε and an operator $S_\varepsilon\in\mathcal{A}(X,Z_\varepsilon)$ such that

$$(*)\qquad \|Tx\| \le \|S_\varepsilon x\| + \varepsilon\cdot\|x\|,\ \forall x\in X.$$

If $\mathcal{A}$ admits a complete ideal quasi-norm, then all Z_ε may be chosen equal to a single Banach space Z and S_ε of the form $S_\varepsilon = N(\varepsilon)\cdot S$ with $S\in\mathcal{A}(X,Z)$ and $N(\varepsilon)>0$; i.e. (*) takes the form

$$(\overset{*}{*})\qquad \|Tx\| \le N(\varepsilon)\cdot\|Sx\| + \varepsilon\cdot\|x\|,\ \forall x\in X,\ \forall\varepsilon>0.$$

The interest in such a result originated from C.Niculescu's paper [11], where he used $(\overset{*}{*})$ with $\mathcal{A}=\mathcal{P}_1$ to define what he proposed to call <u>absolutely continuous operators</u>. So, if AC denotes the ideal of these operators, then

$$AC = \bar{\mathcal{P}}_1^{inj}.$$

This, together with known relations between $\mathcal{K}$, $\mathcal{V}$, and $\mathcal{W}$ implies

$$\mathcal{K} = (AC)^2 \subset AC \subset \mathcal{V}\cap\mathcal{W}.$$

Both inclusions are proper. As for the right one, every operator from ℓ_1 onto a reflexive Banach space which is not super-reflexive provides an example; this will become clear a little later. On the other hand, Rosenthal's ℓ_1-theorem [17] (see also [7]) implies that $AC(X,\cdot)=\mathfrak{K}(X,\cdot)$ holds if and only if X does not contain a copy of ℓ_1.

As $\mathfrak{P}_1$ is the injective hull of the ideal $\mathfrak{I}_1$ of all integral operators, we also may write $AC = \overline{\mathfrak{I}_1}^{inj}$, of course. Actually, in the descriptions of AC, $\mathfrak{P}_1$ and $\mathfrak{I}_1$ can be replaced by $\mathfrak{P}_p$ and $\mathfrak{I}_p$, respectively, for arbitrary $1 \le p < \infty$. This is a consequence of the following important result of Niculescu's [11]. A related but somewhat different proof due to A.Pełczyński may be found in [5].

(2) If K is a compact Hausdorff space, then $\mathfrak{W}(\mathcal{C}(K),\cdot) = AC(\mathcal{C}(K),\cdot)$.

It is obvious that we may replace $\mathcal{C}(K)$ by any abstract $\mathcal{L}_\infty$-space. It was again A.Pełczyński who first noted that $\mathcal{C}(K)$ can even by replaced by an arbitrary subspace X which has the property that every weakly compact subset of X* can be lifted, via the canonical quotient map, to a weakly compact subset of $\mathcal{C}(K)^*$. In fact, the property just described is equivalent to the equation $\mathfrak{W}(X,\cdot) = (\mathfrak{W}\circ\Gamma_\infty)^{inj}(X,\cdot)$. As was shown by F.Delbaen [2] and S.V.Kisliakov [6], certain spaces of analytic functions, like the disc algebra A, belong to this class of Banach spaces. According to [15], however, A fails to be an $\mathcal{L}_\infty$-space.

An appeal to the factorization properties of absolutely p-summing operators now yields the above mentioned identity:

(3) For all $1 \le p < \infty$, $AC = \overline{\mathfrak{P}_p}^{inj}$ $(= \overline{\mathfrak{I}_p}^{inj})$.

4. Some further consequences

It is clear that $\mathfrak{P}_p$ is contained in the ideal BS of all <u>Banach-Saks operators</u>. Recall that $T: X \longrightarrow Y$ is a Banach-Saks operator iff, given any bounded sequence (x_n) in X, we may extract from (Tx_n) a subsequence with norm convergent arithmetic means. Since BS is uniformly closed and injective, $AC \subset BS$ follows. Moreover, $BS \subset \mathfrak{W}$, so that (2) implies:

(4) If K is a compact Hausdorff space, then $\mathfrak{W}(\mathcal{C}(K),\cdot) = BS(\mathcal{C}(K),\cdot)$.

This was first obtained by J.Diestel and C.J.Seifert in [3]. Note also that the remark following (2) applies.

Consider next a Banach spaces X, an $\mathcal{L}_1$-space Y, and a weakly compact operator $T:X \longrightarrow Y$. Let $Q:\ell_1(I) \longrightarrow X$ be a quotient map. By (2) and since $\ell_\infty(I)$ is an injective Banach space, Q^*T^* is the uniform limit of a sequence of 2-factorable (and even of integral) operators, hence the same is true for $(TQ)^{**}$. In particular, T is a Banach-Saks operator. Thus (cf. W.Szlenk [19]):

(5) If Y is a subspace of an $\mathcal{L}_1$-space, then $\mathfrak{W}(\cdot,Y) = BS(\cdot,Y)$.

In the next section, we shall see that (4) and (5) can be considerably strengthened.

5. Ultrapowers

Given Banach spaces X and Y, an operator $T:X \longrightarrow Y$, and an ultrafilter $\mathfrak{U}$ (on some index set), we denote by $X^{\mathfrak{U}}$, $Y^{\mathfrak{U}}$, and $T^{\mathfrak{U}}:X^{\mathfrak{U}} \longrightarrow Y^{\mathfrak{U}}$ the corresponding ultrapowers. Following S.Heinrich [4], we denote, for a given operator ideal $\mathfrak{A}$ and Banach spaces X, Y, by $\mathfrak{A}^{power}(X,Y)$ the set of all operators $T:X \longrightarrow Y$ such that $T^{\mathfrak{U}} \in \mathfrak{A}(X^{\mathfrak{U}},Y^{\mathfrak{U}})$ holds for every ultrafilter $\mathfrak{U}$. In this way, we always obtain an operator ideal,

$$\mathfrak{A}^{power}.$$

It is clear that $\mathfrak{A}^{power}$ is contained in $\mathfrak{A}$. If $\mathfrak{A} = \mathfrak{A}^{power}$, then $\mathfrak{A}$ is called <u>power-stable</u>.

Straightforward arguments lead to

$$\overline{\mathfrak{A}^{power}} \subset (\bar{\mathfrak{A}})^{power}$$

and

$$(\mathfrak{A}^{power})^{inj} \subset (\mathfrak{A}^{inj})^{power}.$$

Thus $\bar{\mathfrak{A}}^{inj}$ is power-stable whenever $\mathfrak{A}$ is. Being ultra-stable [16], $\mathfrak{P}_p$ is a fortiori power-stable. Consequently:

(6) $AC = AC^{power} \subset \mathfrak{W}^{power}$.

The latter inclusion is of course a consequence of the fact that AC consists of weakly compact operators only.

From (2) and (6) we get:

(7) If K is a compact Hausdorff space, then $\mathfrak{W}(\mathcal{C}(K),\cdot) = \mathfrak{W}^{power}(\mathcal{C}(K),\cdot)$.

The argument used to prove (5) also yields:

(8) If Y is (a subspace of) an $\mathcal{L}_1$-space, then $\mathfrak{W}(\cdot, Y) = \mathfrak{W}^{power}(\cdot, Y)$.

For another proof see [1]. (7) and (8) are of course dual to each other.

If $\mathfrak{A}$ is a regular ideal, then $\mathfrak{A}^{power}$ can be described in terms of a certain concept of finite representability of operators, cf. [4]. If $\mathfrak{A}$ is even injective, then a Banach space X satisfies $I_X \in \mathfrak{A}^{power}$ if and only if $I_Y \in \mathfrak{A}$ holds for every Banach space Y which is finitely representable in X in the usual sense. In particular, $I_X \in \mathfrak{W}^{power}$ if and only if X is super-reflexive. More generally, $\mathfrak{W}^{power}$ consists exactly of the uniformly convexifying operators defined by B.Beauzamy [1]; see again [4]. Since uniformly convexifying operators are Banach-Saks operators, (4) and (5) are covered by (7) and (8), respectively.

The above arguments also yields $AC \subset \mathfrak{V}^{power}$. But so far no useful information could be deduced from this fact, the main drawback being that a reasonable description of $\mathfrak{V}^{power}$ seemingly does not exist. All we know is that only finite dimensional Banach spaces X satisfy $I_X \in \mathfrak{V}^{power}$ (the "super-Schur property").

Another immediate consequence of (7) and (8) is the following famous result due to H.P.Rosenthal [18]:

(9) Reflexive quotients (subspaces) of $\mathcal{L}_\infty$-spaces ($\mathcal{L}_1$-spaces) are super-reflexive.

By using cotype arguments [10] together with factorization properties of absolutely p-summing operators, we see that every reflexive quotient of an $\mathcal{L}_\infty$-space is in fact a quotient of some space $\mathcal{L}_q(\mu)$, $2 \leq q < \infty$. Dually, every reflexive subspace of an $\mathcal{L}_1$-space is a subspace of some $\mathcal{L}_p(\nu)$, $1 < p \leq 2$.

Of course, (7) through (9) can be modified by taking into account the remark following (2).

6. A classification of absolutely continuous operators

Let $\mathfrak{A}$ be a complete quasi-normed operator ideal. According to (1), an operator $T: X \longrightarrow Y$ belongs to $\mathfrak{A}^{inj}$ if and only if there is a Banach space

Z, an operator $S \in \mathcal{A}(X,Z)$, and a function N of $]0,\infty[$ into itself such that

$$(\alpha) \quad \|Tx\| \le N(\varepsilon)\cdot\|Sx\| + \varepsilon\cdot\|x\|, \ \forall x\in X, \ \forall \varepsilon>0.$$

We ask for the impact of controlling the growth of N considered as a function of $\frac{1}{\varepsilon}$. More precisely, we ask what it means to require that

$$(\beta) \quad \|Tx\| \le \varepsilon^{-r}\cdot\|Sx\| + \varepsilon\cdot\|x\|, \ \forall x\in X, \ \forall \varepsilon>0$$

holds for some $r>0$ and $S\in\mathcal{A}(X,Z)$. By calculating the minimum of the function of ε appearing on the right hand side of (β), for $x\in X$ fixed, we see that (β) is equivalent to the "interpolation formula"

$$(\gamma) \quad \|Tx\| \le \|\tilde{S}x\|^{1-\Theta}\cdot\|x\|^{\Theta}, \ \forall x\in X,$$

where $\Theta := r\cdot(r+1)^{-1}$ and $\tilde{S}\in\mathcal{A}(X,Z)$ is just a constant multiple of the operator S occurring in (β).

The operators determined by (γ) form an injective complete quasi-normed ideal which we denote by

$$\mathcal{A}_\Theta.$$

If $\mathcal{A}$ is even a Banach ideal, then so is $\mathcal{A}_\Theta$. Also, $\mathcal{A}_\Theta$ is power-stable whenever $\mathcal{A}$ is.

Obviously,

$$\mathcal{A}^{inj} \subset \mathcal{A}_{\Theta_1} \subset \mathcal{A}_{\Theta_2} \subset \bar{\mathcal{A}}^{inj}$$

if $0<\Theta_1<\Theta_2<1$. We shall see below that these inclusions are proper. We do not intend to discuss the interpolative aspects of the construction just described in this note. More details on these and related questions can be found in [8]. Here we only report on some results concerning the case $\mathcal{A}=\mathcal{P}_p$. From what was said before,

$$\mathcal{P}_p \subset (\mathcal{P}_p)_\Theta \subset AC .$$

The left inclusion can be improved:

(10) $\mathcal{P}_{\frac{p}{1-\Theta}} \subset (\mathcal{P}_p)_\Theta$, for all $1\le p<\infty$ and $0<\Theta<1$.

This inclusion is proper in general. To see this, let $\mathcal{S}_r$ denote the r-th Schatten - von Neumann class. By [14], $\mathcal{P}_{\frac{p}{1-\Theta}}(\mathcal{L}_2,\mathcal{L}_2)$ is precisely the space $\mathcal{S}_2(\mathcal{L}_2,\mathcal{L}_2)$ of Hilbert-Schmidt operators. On the other hand, the following can be shown:

(11) If $1\le p\le 2$, then $(\mathcal{P}_p)_\Theta(\mathcal{L}_2,\mathcal{L}_2) = \mathcal{S}_{\frac{2}{1-\Theta}}(\mathcal{L}_2,\mathcal{L}_2)$, $\forall 0<\Theta<1$.

Note that $(\mathcal{P}_p)_\Theta(\mathcal{L}_2,\cdot) = (\mathcal{P}_2)_\Theta(\mathcal{L}_2,\cdot)$ if $1\le p\le 2$.

Since $\bigcup_{p>0} \ell_p$ is a proper subset of c_o and since $AC(X,\cdot) = \mathfrak{K}(X,\cdot)$ holds e.g. if X is reflexive, we see that $\bigcup_{0<\Theta<1} (\mathfrak{P}_2)_\Theta$ is properly contained in AC.

Next we sketch how to prove that the inclusion in (10) is also best possible, in the sense that $\mathfrak{P}_r \not\subset (\mathfrak{P}_p)_\Theta$ if $r > \frac{p}{1-\Theta}$. The key result is:

(12) $(\mathfrak{P}_p)_\Theta \subset \mathfrak{P}_{\frac{p}{1-\Theta},p}$ for $1 \leq p < \infty$ and $0<\Theta<1$.

As in [16], $\mathfrak{P}_{\frac{p}{1-\Theta},p}$ denotes the ideal of $(\frac{p}{1-\Theta},p)$-summing operators. It follows again from (11) that this inclusion is strict if $1 \leq p < 2$. In fact (cf. [16], 22.1.12), if $\Theta < \frac{1}{2}p$, then $\mathfrak{P}_{\frac{p}{1-\Theta},p}(\mathcal{L}_2,\mathcal{L}_2) = \mathcal{S}_{\frac{2p}{p-2\Theta}}(\mathcal{L}_2,\mathcal{L}_2)$, whereas $\mathfrak{P}_{\frac{p}{1-\Theta},p}(\mathcal{L}_2,\mathcal{L}_2) = \mathcal{L}(\mathcal{L}_2,\mathcal{L}_2)$ if $\Theta \geq \frac{1}{2}p$.

Combining (11) and (12) with results of B.Maurey [9] , we get:

(13) If X is an $\mathcal{L}_\infty$-space, then $\mathfrak{P}_{\frac{p}{1-\Theta}}(X,\cdot) \subset (\mathfrak{P}_p)_\Theta(X,\cdot) \subset \bigcap_{\varepsilon>0} \mathfrak{P}_{\frac{p}{1-\Theta}+\varepsilon}(X,\cdot)$.

It would be interesting to know which of these inclusions are proper. We also would like to know if (13) could lead to any improvements of the known characterizations [10] of the cotypes of a Banach space.

7. Concluding remarks

In contrast to the situation for absolutely p-summing operators, it is not true that for every $S \in (\mathfrak{P}_p)_\Theta(X,Y)$ there is a Banach space $\tilde{Y}$, containing Y as a subspace, and an operator $\tilde{S} \in (\mathfrak{P}_p)_\Theta(\mathcal{C}(B_{X^*}),\tilde{Y})$ extending S. This follows e.g. from (11) and (13). It is not clear what the corresponding situation is for the ideal AC:

Problem 1 ([5]): Given $S \in AC(X,Y)$, does there exist a Banach space $\tilde{Y}$ containing Y and a weakly compact extension $\tilde{S}:\mathcal{C}(B_{X^*}) \longrightarrow \tilde{Y}$ of S ?

Note that such an operator S admits a "Riesz representation" by the $\tilde{Y}$-valued measure on B_{X^*} defined by $\tilde{S}$. We may reformulate problem 1 by asking if AC and $(\mathfrak{W}\circ\Gamma_\infty)^{inj}$ coincide. By definition, AC is uniformly closed. Is that also true for $(\mathfrak{W}\circ\Gamma_\infty)^{inj}$?

As we have already mentioned, $AC(X,\cdot)=\mathfrak{K}(X,\cdot)$ is equivalent to saying that X does not contain a copy of ℓ_1. Let $\mathcal{H}$ be the class of all Banach spaces X such that all operators from ℓ_1 (or any infinite dimensional $\mathcal{L}_1$-space) into X are absolutely continuous. Every Banach space in $\mathcal{H}$ is super-reflexive. Further, belonging to $\mathcal{H}$ is a super-property for Banach spaces. $\mathcal{H}$ is stable with respect to the formation of subspaces, quotients, and duals, and it contains all the $\mathcal{L}_p$-spaces, $1<p<\infty$. Correspondingly, we may say that X belongs to $\mathcal{H}$ if and only if $\mathcal{L}(\ell_1,X) = \overline{\Gamma_p}^{inj}(\ell_1,X)$ for some (and hence all) $1<p<\infty$.

Problem 2: Do the Schatten- von Neumann classes $\mathcal{S}_p(\mathcal{L}_2,\mathcal{L}_2)$ for $1<p<\infty$ belong to $\mathcal{H}$?

We do not even know if there are super-reflexive Banach spaces which do not belong to $\mathcal{H}$.

Some more information is available for Banach spaces X such that $\mathcal{L}(\ell_1,X) = (\mathcal{P}_p)_\Theta(\ell_1,X)$ for some fixed $1\leq p<\infty$ and $0<\Theta<1$. Again, this defines a super-property for Banach spaces, and it allows to give lower bounds (upper bounds) for the supremum (infimum) of the possible types (cotypes) of the Banach space in question. Details may be found in [8].

Acknowledgement

The authors wish to thank A.Pietsch and G.Pisier for several stimulating discussions.

References

[1] Beauzamy,B.: Opérateurs uniformément convexifiants. Studia Math. 57 (1976) 103-139.

[2] Delbaen,F.: Weakly compact operators on the disk algebra. Journ. of Algebra 45 (1977) 284-294.

[3] Diestel,J.; Seifert,C.J.: The Banach-Saks ideal, I. Operators acting on $\mathcal{C}(\Omega)$. Comment.Math., Tom.spec.hon.L.Orlicz (1979) 109-118 and 343-344.

[4] Heinrich,S.: Finite representability and super-ideals of operators. Dissertationes math. CLXII (1980).

[5] Jarchow,H.: Locally Convex Spaces. Stuttgart 1981.

[6] Kisliakov,S.V.: On the conditions of Dunford-Pettis, Pełczyński, and Grothendieck. Dokl.Akad.Nauk SSSR 225 (1975) 1252-1255 (Engl. transl. in Soviet Math.Dokl. 16 (1975) 1616-1620).

[7] Lindenstrauss,J.; Tzafriri,L.: Classical Banach Spaces I, II. Berlin- Heidelberg- New York 1977, 1979.

[8] Matter,U.: Thesis, Universität Zürich (1985).

[9] Maurey,B.: Une nouvelle characterization des applications (p,q)-sommantes. École Polytechn.Paris, Sém.Maurey-Schwartz 1973-1974, exp.no. 12.

[10] Maurey,B.; Pisier,G.: Séries de variables aléatoires vectorielles indépendentes et propriétés géométriques des espaces de Banach. Studia Math. 58 (1976) 45-90.

[11] Niculescu,C.: Absolute continuity in Banach space theory. Rev.Roum. Pures Appl. 24 (1979) 413-422.

[12] Pełczyński,A.: Banach spaces on which every unconditionally converging operator is weakly compact. Bull.Acad.Polon.Sci., Sér.Sci. Math.Astron.Phys., 10 (1962) 641-648.

[13] ----- " -----: On strictly singular and strictly cosingular operators. Bull.Acad.Polon.Sci., Sér.Sci.Math.Astron.Phys., 13 (1965) 31-36 and 37-41.

[14] ----- " -----: A characterization of Hilbert-Schmidt operators. Studia Math. 28 (1966/67) 355-360.

[15] ----- " -----: Sur certaines propriétés isomorphiques nouvelles des espaces de Banach de fonctions holomorphes A et H^{∞}. C.R.Acad.Sci. Paris A 279 (1974) 9-12.

[16] Pietsch,A.: Operator ideals. Berlin 1978; Amsterdam-Oxford-New York 1980.

[17] Rosenthal,H.P.: A characterization of Banach spaces containing ℓ^1. Proc.Nat.Acad.Sci.USA 71 (1974) 2411-2413.

[18] ----- " ------: Some applications of p-summing operators to Banach space theory. Studia Math. 58 (1976) 21-43.

[19] Szlenk,W.: Sur les suites faiblement convergentes dans l'espace L. Studia Math. 25 (1965) 337-341.

A NOTE ON TOEPLITZ OPERATORS

D. Khavinson*
Department of Mathematical Sciences
University of Arkansas
Fayetteville, Arkansas 72701

1. Introduction

Let H be a Hilbert space with an inner product $\langle,\rangle : H \times H \to \mathbb{C}$. Let $T:H \to H$ be a bounded linear operator. We use the standard notation $sp(T)$ for the spectrum of T, i.e. for the set $\{\lambda \in \mathbb{C} : T - \lambda I$ is not invertible$\}$. Here, $I:H \to H$ is the identity operator. As usual, the adjoint operator of T will be denoted by T^*. We recall that an operator $S:H \to H$ is called positive $(S \geq 0)$ if $(Sx,x) \geq 0$ for all $x \in H$. If T is not a normal operator, i.e. $T^*T \neq TT^*$, then the quantity $\|T^*T-TT^*\| \neq 0$ can be viewed as a measure of the "abnormality" of T. The operator T^*T-TT^* is usually denoted by $[T^*,T]$ and called the self-commutator of T.

In [8] Putnam proved the following theorem.

If $[T^*,T] \geq 0$, then

$$\|[T^*,T]\| \leq \frac{(\text{area}(sp(T))}{\pi} \tag{1}$$

This result turned out to be very useful for many problems in operator theory and in function theory. In particular, many interesting applications to the problems in the analytic function theory in the unit disk and in the unit ball are contained in [3]. Also, [3] contains many references concerning various corollaries from the Putnam theorem.

The main purpose of this paper is to show that for a wide class of subnormal operators T, so-called Toeplitz operators with an analytic symbol, there exists a natural lower bound for $\|[T^*,T]\|$. This lower bound is also related to the geometry of the $sp(T)$. The corresponding statement, its proof, and some corollaries are contained in §2. We remark that as one of the most surprising corollaries, we obtain the classical isoperimetric inequality with sharp constants.

§3 contains further generalizations to the case of uniform algebras.

---

*This work was supported in part by National Science Foundation EPSCOR Grant No. ISP-8011447.

§2. Toeplitz Operators with an Analytic Symbol

Let G be a finitely connected domain in $\mathbb{C}$ with the boundary Γ consisting of n smooth closed Jordan curves. $R(\overline{G})$ denotes the uniform closure of rational functions with poles outside of $\overline{G}$. The Smirnov class $E_2 = E_2(G)$ (cf. [6], Ch. 10) is defined as the closure of $R(\overline{G})$ in $L^2(ds)$. Here, ds is the Lebesgue measure on Γ. $P(G) = \int ds$ denotes the perimeter of G. Let ϕ be a function analytic in a neighborhood of $\overline{G}$. Define an operator $T_\phi : E_2 \to E_2$ by $T_\phi f = \phi \cdot f$.

THEOREM 1.

$$\|[T_\phi^*, T_\phi]\| \geq \frac{4\ \mathrm{area}^2(\mathrm{sp}(T))}{\|\phi'\|_{E_2}^2 \cdot P(G)} . \tag{2}$$

Proof. Let $P: L^2(\Gamma) \to E_2$ be the orthogonal projection. The following argument is standard (cf. [3]). For any $g, h \in E_2$, we have

$$\begin{aligned} \langle T_\phi^* g, h\rangle &= \langle g, T_\phi h\rangle = \langle g, \phi h\rangle = \\ &= \langle \overline{\phi} g, h\rangle = \langle P(\overline{\phi} g), h\rangle \end{aligned}$$

Hence, $T_\phi^* g = P(\overline{\phi} g)$. Since $[T_\phi^*, T_\phi]$ is a normal operator on E_2, then (see [9], Th. 12.25)

$$\|[T_\phi^*, T_\phi]\| = \sup_{\substack{g \in E_2 \\ \|g\|_{E_2}}} \langle [T_\phi^*, T_\phi] g, g\rangle. \tag{3}$$

Fix $g \in E_2$, $\|g\| = 1$. We have

$$\begin{aligned} \langle (T_\phi^* T_\phi - T_\phi T_\phi^*) g, g\rangle &= \|T_\phi g\|^2 - \|T_\phi^* g\|^2 = \|\phi g\|^2 - \|P(\overline{\phi} g)\|^2 = \\ &= \|\overline{\phi} g\|_{L_2}^2 - \|P(\overline{\phi} g)\|_{E_2}^2 = \left[\mathrm{dist}_{L^2}(\overline{\phi} g, E_2)\right]^2. \end{aligned}$$

From (3) we obtain that

$$\|[T_\phi^*, T_\phi]\| = \sup_{\|g\|_{E_2}=1} \{\inf_{f \in E_2} \|\overline{\phi} g - f\|_{L_2}\}^2. \tag{4}$$

Taking $g = \frac{1}{\sqrt{P(G)}}$, we find

$$\|[T_\phi^*, T_\phi]\| \geq \{\inf_{f \in E_2} \|\overline{\phi} - f\|_{L^2}\}^2 \cdot \frac{1}{P(G)} . \tag{5}$$

Fix $f \in E_2$. Then, from the standard duality argument based on the Hahn-Banach Theorem (see [7], Ch. II) we have

$$\|\bar{\phi}-f\|_{L^2} = \sup_{\substack{g \in L^2 \\ \|g\|_{L^2}=1}} \{|\int_\Gamma (\bar{\phi}-f)\bar{g}ds|\}.$$

Taking $g(\zeta)\Big|_\Gamma = \frac{d\bar{\zeta}}{ds} \cdot \overline{\phi'(\zeta)} \cdot \frac{1}{\|\phi'\|_{E_2}}$, we obtain

$$\|\bar{\phi}-f\|_{L^2} \geq |\int_\Gamma (\bar{\phi}-f)\phi' d\zeta| \cdot \frac{1}{\|\phi'\|_{E_2}} =$$

$$= \frac{1}{\|\phi'\|_{E_2}} |\int_\Gamma \bar{\phi} \cdot \phi' d\zeta| \qquad (6)$$

($\int_\Gamma f\phi' d\zeta = 0$, since $f \in E_2$, $\phi' \in E_2$ and therefore we can apply Cauchy's Theorem). But, according to Stokes' Theorem,

$$|\int_\Gamma \bar{\phi}\phi' d\zeta| = |2i \iint_G |\phi'|^2 dxdy| = |2i(\text{area } \phi(G) \text{ with multiplicity})| \geq$$

$$\geq 2 \text{ area } (\phi(G)). \qquad (7)$$

Since $sp(T_\phi) = \overline{\phi(G)}$ (see [5], Ch. II), combining (5), (6), and (7), we complete the proof.

Take $\phi \equiv z$ (the shift operator). Then $sp(T_z) = \bar{G}$, $\|\phi'\|^2_{E_2} = P(G)$. Thus, we obtain the following

COROLLARY 1.

$$\frac{\text{area}(G)}{\pi} \geq \|[T_z^*, T_z]\| \geq 4 \frac{\text{area}^2(G)}{P^2(G)}$$

In particular,

$$P^2(G) \geq 4\pi \text{area}(G).$$

This is the classical isoperimetric inequality with sharp constants. (It becomes equality for the disks.)

Combining (1), (2), and (7), and applying a standard approximation argument, one obtains the following.

COROLLARY 2. *For all functions* f *analytic in* G, *the following inequality holds:*

$$\iint_G |f|^2 dxdy \le \frac{P(G)}{4\pi} \|f\|^2_{E_2}.$$

COROLLARY 3. *If* ϕ *is proper, then*

$$n = \textit{multiplicity of } \phi \le \frac{1}{2}\sqrt{\frac{P(G)}{\pi \operatorname{area}(\phi(G))}}\, \|\phi'\|_{E_2}.$$

REMARK. We want to point out that the equality (4) is itself quite important. Consider the following situation. Let $X \subset \mathbb{C}$ be a compact set and let μ be a positive Borel measure on X. Denote by $H_2(\mu)$ the closure of all polynomials in $L^2(\mu)$. Let $T_z : H_2 \to H_2$ be defined by $H_2 \ni f \to zf$. The same calculation as in the proof of Theorem 1 yields

$$\|[T_z^*, T_z]\| = \sup_{\substack{g \in H_2(\mu) \\ \|g\|=1}} \{\inf_{f \in H_2(\mu)} \|\bar{z}g - f\|_{L^2(\mu)}\}^2.$$

If T_z is normal, then taking $g = \sqrt{\frac{1}{\|\mu\|}}$, we obtain that $\bar{z} \in H_2(\mu)$. Moreover, if T_z is normal, then all $(T_z)^n$, $n = 1,2,\ldots$ are normal Taking for a fixed n $g_m^n = \frac{z^m}{\|z^m\|_2}$, $m = 0,1,\ldots,$ and applying (4) to T_z^n, we obtain that all monomials $\bar{z}^n z^m \in H_2(\mu)$, $n = 1,2,\ldots,$ $m = 0,1,\ldots$. Then, the Stone-Weierstrass Theorem yields that $H_2(\mu) \supset C(X)$ and, therefore, $H_2(\mu) = L^2(\mu)$. This fact is known (see [4]), but we have not been able to locate a proof in the literature.

3. Toeplitz Operators on Banach Algebras

Let A be a uniform algebra with maximal ideal space M. Let $x \in M$ and let m be a (positive) representing measure for x supported on M. $sp(f) = \{\lambda \in \mathbb{C} : f - \lambda$ is not invertible in $A\}$. [For the basic account on Banach algebras, see [7].) Let $H_2(m)$ stand for the closure of A in $L^2(m)$. For any $f \in A$, define a Toeplitz operator $T_f : H_2(m) \to H_2(m)$ by $T_f(g) = f \cdot g$.

THEOREM 2.

$$\frac{\operatorname{area}(sp(f))}{\pi} \ge \|[T_f^*, T_f]\| \ge \int |f - f(x)|^2 dm. \tag{8}$$

Proof. The first inequality is (8) is a direct corollary from Putnam's Theorem. Without loss of generality we can assume that $f(x) = 0$. The same argument as in the proof of Theorem 1 yields

$$\|[T_f^*, T_f]\| = \sup_{\substack{g \in H_2(m) \\ \|g\|_{H_2}=1}} \left[\operatorname{dist}_{L^2(m)}(\bar{f}g, H_2)\right]^2 \geq$$

$$\geq \left[\operatorname{dist}_{L^2(m)}(\bar{f}, H_2)\right]^2. \tag{9}$$

Fix $\phi \in H_2(m)$. Then, by duality

$$\|\bar{f}-\phi\|_{L^2} = \sup_{\substack{g \in L^2(m) \\ \|g\|_{L^2}=1}} \left|\int (\bar{f}-\phi)\bar{g}\,dm\right|.$$

Taking $g = \frac{\bar{f}}{\|f\|_{H_2(m)}}$ and using the fact that m is a representing measure for x, we obtain

$$\|\bar{f}-\phi\|_{L^2} \geq \int |f|^2 dm. \tag{10}$$

From (9) and (10) our statement follows.

REMARK. The inequality

$$\frac{\operatorname{area}(\operatorname{sp}(f))}{\pi} \geq \int |f-f(x)|^2 dm$$

is due to H. Alexander (see [1]). Also see [2], [3] for further discussions and applications to function theory.

COROLLARY 4. *If* T_f *is normal, then* $f \equiv \text{const}$ *on* $\operatorname{supp} m$.

COROLLARY 5. *Let* m *be supported on the Šilov boundary* Γ *of* A *and let* $f(x) = 0$, $|f|\big|_\Gamma \equiv 1$ *for* $f \in A$. *Then* $\operatorname{sp}(f) = \{z : |z| \leq 1\}$ *and* $\|[T_f^*, T_f]\| = 1$.

References

1. H. Alexander, On the area of the spectrum of an element of a uniform algebra, Complex Approximation, Proceedings, Quebec, July 3-8, 1978,

ed. by B. Aupetit, Birkhauser (1980), 3-12.
2. H. Alexander, B. A. Taylor, and J. L. Ullman, Areas of projections of analytic sets, Invent. Math. 16(1972), 335-341.
3. S. Axler and J. H. Shapiro, Putnam's Theorem, Alexander's spectral area estimate and VMO, 1983 (preprint).
4. J. Brennan, Invariant subspaces and subnormal operators, Proc. of Symposia in Pure Math., Vol. XXXV, Part 1(1979), 303-309.
5. R. Douglas, Banach algebra techniques in operator theory, Pure and Appl. Mathematics, Vol. 49, Academic Press, 1972.
6. P. Duren, Theory of H_p-spaces, Academic Press, New York, 1970.
7. T. Gamelin, Uniform algebras, Prentice-Hall, 1969.
8. C. R. Putnam, An inequality for the area of hyponormal spectra, Math. Zeit. 116(1970), 323-330.
9. W. Rudin, Functional Analysis, McGraw-Hill, 1973.

On the (F) Property

by
Thomas A. Metzger

§1. Introduction: Let Δ be the unit disk in the complex plane and define the Hardy classes $H^p(\Delta)$ $(1 \leq p \leq \infty)$ to be the Banach space of holomorphic functions $f(z)$ with the norm

$$||f||_p^p = \lim_{r \to 1} \frac{1}{2\pi} \int_0^{2\pi} |f(re^{i\theta})|^p d\theta, \ (1 \leq p < \infty)$$

$$||f||_\infty = \sup_{z \in \Delta} |f(z)| \ , \quad p = \infty.$$

As is well known any such f can be factored as f = FI where I is the inner factor and F is the outer factor of f. These spaces have been the object of many papers, one can see the recent text of Koosis [K] for many of the properties of these spaces of functions. In this work we considered property (F) which was originally introduced by Havin [H] in a report on earlier work by the Russian school.

Definition 1: A Banach subspace X of H^1 is said to have the (F) property if, given any h in X and any inner function I which divides h (i.e., $h/I \in H^1(\Delta)$) then $h/I \in X$.

There are many subspaces of $H^1(\Delta)$ which have property (F), for example, $H^p(\Delta)$ themselves, $A = \{f : f \text{ is continuous on } \bar{\Delta} \text{ and holomorphic on } \Delta\}$, $H_n^p = \{f : f^{(n)} \in H^p(\Delta)\}$, $Lip(\alpha,n) = \{f \in H^\infty(\Delta) : |f^{(n)}(z) - f^{(n)}(\zeta)| \leq C\,|z-\zeta|^\alpha\}$, and BMOA the space of analytic functions with bounded mean oscillation. The proofs of these results appear in a variety of sources. For references one can check the recent work of Shirokov [S1 and S2]. It is also known that some subspaces of $H^1(\Delta)$ do not possess the (F) property. Those include $\ell^p = \{f(z) = \sum_{n=0}^{\infty} a_n z^n : \{a_n\} \in \ell^p\}$ for $1 \leq p < 2$, many weighted ℓ^p spaces and the subalgebra $H^\infty(\Delta) \cap B_0$ where B_0 is the little Bloch space, i.e., the space of holomorphic f such that

$$(1 - |z|^2)\ |f'(z)| \to 0 \text{ as } |z| \to 1.$$

(See [A] and [S2]). Based on these examples Shirokov, in [S2], assets "Spaces without property (F) usually appear when the norm $||\cdot||_X$ is only indirectly connected with the values of the function f". Thus, it would seem useful to give the following two examples of spaces which do not satisfy the (F) property and yet $||f||_X$ can be taken to be the usual Hardy space norm so that the norm depends upon the value of the functions involved.

Theorem 1: There exists a Banach space X_1 such that if f belongs to X_1 and I is a non-constant inner function which divides f then f/I does not belong to X_1.

Theorem 2: There exists a Banach space X_2 such that if f belongs to X_2 and I is an inner function which divides f then f/I belongs to X_2 if and only if I belongs to X_2.

It is worth noting that in both of these examples if $I \in X$ then $f/I \in X$, assuming that $f \in X$ and I divides f. This will be denoted as the weak (F) property and in §3 it will be shown that ℓ^1 satisfies the weak (F) property.

§2. Construction of Examples: Both of the examples are based on Hardy classes on Riemann surfaces, let W be a Riemann surface which possesses a Green's function and let Γ be the Fuchsian group which uniformizes W, i.e., $W \cong \Delta/\Gamma$. If $\pi : \Delta \to W$ is the universal covering map then any holomorphic function F on W can be "pulled back" to a function f holomorphic on Δ by the equation

$$f(z) = F(\pi(z)) \text{ for all } z \text{ in } \Delta.$$

This pulled back function f will be automorphic with respect to Γ, i.e.,

$$(f\circ\gamma) = f \text{ for all } \gamma \text{ in } \Gamma. \tag{1}$$

One can then define a subspace of $H^p(\Delta)$ as the space of automorphic functions in $H^p(\Delta)$, i.e,

$$H^p(\Delta/\Gamma) = \{f \in H^p : (1) \text{ holds}\}.$$

It is well known (see [R]) that $H^p(\Delta/\Gamma)$ is isometrically isomorphic to $H^p(W)$ the Hardy space on the associated Riemann surface.

With these preliminaries the construction of the examples is simple. To prove Theorem 1 one needs only construct a Riemann surface W_1 such that $H^\infty(W_1)$ consists only of constants and $H^p(W_1)$ contains non-constant functions for all $p \le p_0 < \infty$. Thus if $f \in H^p(\Delta/\Gamma_1)$ for $p \le p_0$, where Γ_1 is the Fuchsian group uniformizing W_1, and I is an inner function which divides f then the fact that f/I is automorphic implies that I is automorphic and so I is a constant and Theorem 1 is proved.

The proof of Theorem 2 is similar, one merely constructs a Riemann surface W_2 and an associated Fuchsian group Γ_2 such that $H^\infty(W_2)$ contains non-constant functions. Then for X_2 one can take $H^p(\Delta/\Gamma_2)$ for any p in $[1,\infty]$. Again if $f \in X_2$ and f/I belongs to X_2 then I must be automorphic and so $I \in X_2$ and the converse is trivial.

§3. Weak (F) Property: In both of the examples given above the spaces satisfy:

Definition 2: A Banach subspace X in $H^1(\Delta)$ is said to have the weak (F) property if given any f in X and any inner function I which both divides f and also belongs to X then f/I belongs to X.

The fact that this is not just a property of the spaces $H^p(\Delta/\Gamma)$ is the point of

Proposition 3: ℓ^1 satisfies the weak (F) property.

Proof: We first note that if I is an inner function and I belongs to ℓ^1 then I is a finite Blaschke product. Thus it suffices to prove the assertion for $I(z) = {}^{z-a}/(1-\bar{a}z)$ for some $a \in \Delta$. Since f/I is holomorphic it follows that $f((a)) = 0$ and so $f(z) = (z-a)g(z)$ where $g(z)$ has ℓ^1 Taylor coefficients. Thus, $f/I = (1-\bar{a}z)g(z)$ which is clearly an element of ℓ^1 and the proof is complete.

It would be of interest to see if the other known examples of spaces which do not satisfy the (F) property possess the weak (F) property. The author hopes to return to this topic in the near future. Finally, as was noted by D. Khavinson, not all subspaces of $H^1(\Delta)$ possess the weak (F) property. The simplest example results by considering $X = IH^\infty$, where I is an inner function. This X is a closed subalgebra of H^∞ but division by I itself shows it does not satisfy the weak (F) property.

References

[A] J. M. Anderson, "On division by inner functions" Comment Math. Helv. 54 (1979), 309-317.

[Ha] M. Hasumi, Hardy Classes on Infinitely Connected Riemann Surfaces, Lecture Notes in Mathematics, Vol. 1027 Springer Verlag Berlin 1983.

[H] V. P. Havin, "On the factorization of analytic functions smooth up to the boundary" (in Russian) Zap Nauch Sem LOMI 22 (1971), 202-205.

[M.H.] M. Heins, Hardy Classes on Riemann Surfaces. Lecture Notes in Mathematics, Vol. 98 Springer-Verlag Berlin 1968.

[D.H.] D. Hejhal, "Classification Theory for Hardy classes of Analytic Functions" Ann. Acad. Sci. Fenn. Ser. AI no. 566 (1973) 1-28.

[S.K.] S. Kobayaski, "On a classification of plane domains for Hardy classes" Proc. Amer. Math. Soc. 68 (1978) 79-82.

[K] P. Koosis, Introduction to H^p Spaces London Math Soc. Lecture Notes Series No 40. Cambridge U Press, Cambridge, 1980.

[R] W. Rudin, "Analytic Functions of Class H^p" Trans. Amer. Math. Soc. 78 (1955) 46-66.

[S1] N. A. Shirokov, "Division and Multiplication by Inner Fucntions in Spaces of Analytic Fucntions Smooth up to Boundary" Complex Analysis and Spectral Theory, Lecture Notes in Mathematics № 864, Springer Verlag Berlin (1981) pp. 413-439.

[S2] N. A. Shirokov, "Division by an Inner Function Does Not Change the Class of Smoothness" Soviet Math Dokl 27 (1983) 191-193.

Volume Approach and Iteration Procedures

in Local Theory of Normed Spaces

V.D. Milman

Tel Aviv University

et

I.H.E.S. (France)

We analyze in this paper one way to obtain results valid for arbitrary finite dimensional normed spaces using volume computation in $\mathbb{R}^n$. There are two standard ways of using volume in Local Theory (the theory about asymptotic properties of finite dimensional normed spaces). One is to construct norms with special non-trivial properties (see, e.g. [G]). The other one uses volume computations to find nice properties of given special normed spaces, e.g. ℓ_p^n, $1 \leq p < 2$ (see [K],[Sz]). Naturally, one may also consider a normed space satisfying volume inequalities previously observed for ℓ_1^n or prove these inequalities for a special family of spaces to obtain the same type of results (see [Sz T]). However, it was shown in $[M_1]$ and $[M_2]$ that for arbitrary normed finite dimensional spaces one may obtain general results using Szarek's volume approach (see [Sz]) as an intermediate step. At the same time we don't have any conditions connected with volume and results are not connected with volume either. We explicitly extract this application of volume in Proposition 2.1 and present several consequences.

Another idea exploited in this paper is an iteration procedure (as in $[M_2]$) to eliminate some logarithmic factors which otherwise often appear in Local Theory.

1. Notations. Let X be an n-dimensional normed space, i.e. $\mathbb{R}^n$ with the norm $\|\cdot\|$, and let (x,y) be an inner product on X and consequently $|x| = (x,x)^{1/2}$ be the Euclidean norm on X. Then the dual norm $\|\cdot\|^*$ is naturally defined by

$$\|x\|^* = \sup_{y \neq 0} \frac{|(x,y)|}{\|y\|} .$$

If a and b are positive numbers such that for any $x \in X$

$$\frac{1}{a}|x| \leq \|x\| \leq b|x|$$

then, clearly

$$\frac{1}{b}|x| \leq \|x\|^* \leq a|x| .$$

Let K, K^* and D be the unit balls of the spaces X, $X^* = (\mathbb{R}^n, \|.\|^*)$ and Euclidean space $(\mathbb{R}^n, |.|)$ respectively. For example, $K = \{x \in \mathbb{R}^n : \|x\| \leq 1\}$. In some of the constructions below we begin with a convex symmetric compact body $A \subset \mathbb{R}^n$ and define the norm $\|\cdot\|_A$ such that A is the unit ball of $(\mathbb{R}^n, \|\cdot\|_A)$. Denote $M(X) \equiv M_K = \int_{x \in S^{n-1}} \|x\| \, d\mu(x)$, where $S^{n-1} = \{x \in X : |x| = 1\}$ and $\mu(x)$ is the normalized invariant measure on S^{n-1}. Similarly $M(X^*) \equiv M_{K^*} = \int_{x \in S^{n-1}} \|x\|^* d\mu(x)$. We use also a usual (n-dimensional) Lebesgue measure (Vol) on $\mathbb{R}^n$ normalized so that the induced measure on S^{n-1} coincides with $\mu(x)$.

As usual, $T_2(x)$, $C_2(x)$ and Rad_X are the type 2-constant, the cotype 2-constant and the norm of Rademacher projection of space X respectively. These notions were first studied in [MaP] (see, e.g. [M. Sch]). The Banach-Mazur distance $d(X, \ell_2^n)$ between the space X and ℓ_2^n is often denoted d_X.

2. Use of volume.

2.1 Proposition. Let $\frac{1}{a}|x| \le \|x\| \le b|x|$ for any $x \in X$ and let $M(X^*) \le 1$. Then $\forall \lambda < 1$ there exists a subspace $E_\lambda \subset X$ such that $\dim E_\lambda \ge \lambda n$ and for any $x \in E_\lambda$

$$\frac{1}{c(\lambda)}|x| \le \|x\| \le b|x| \qquad ,$$

where $c(\lambda) = 2(4\pi)^{1/(1-\lambda)}$.

Proof : By Urysohn inequality ([U] ; see, e.g. $[M_1]$)

$$\left(\frac{\mathrm{vol}\ K}{\mathrm{vol}\ D}\right)^{1/n} \le M_{K^*} \qquad .$$

We use this inequality for the convex body $A = \mathrm{Conv}(D \cup K)$ instead of K. Then $\|x\|_{A^*} = \max(\|x\|^*, |x|)$,

$$(2.2) \qquad (\mathrm{vol}\ A/\mathrm{vol}\ D)^{1/n} \le M_{A^*} \le 2$$

and the norm $\|\cdot\|_A$ satisfies

$$(2.3) \qquad \frac{1}{a}|x| \le \|x\|_A \le |x|$$

(also clearly $\|x\|_A \le \|x\|$). By Szarek result [Sz] (2.2) and (2.3) imply for any $\lambda < 1$ the existence of a subspace $E_\lambda \subset X$, $\dim E \ge \lambda n$, such that for any $x \in E_\lambda$

$$\frac{1}{c(\lambda)}|x| \le \|x\|_A \qquad \square$$

2.4 Remark. It is known (and easy to show) that $M_{K \cap E_\lambda} \le \sqrt{1/\lambda}\, M_K$ for any subspace E_λ, $\dim E_\lambda \ge \lambda n$.

The following corollaries of the above Proposition we obtain using special inner products on X.

3. Subsequent results of Lewis [L]- Figiel-Tomĉzak [FT] - Pisier [P] were exploited in $[M_2]$ to construct an inner product on X such that $M_{K^*} \leq 1$ and $M_K \leq c\ell n(d_X+1)$, where c is an absolute constant and d_X denotes the Banach-Mazur distance between X and ℓ_2^n . We apply Proposition 2.1 to find a subspace E_λ . Then for any $x \in E_\lambda^*$

$$\frac{1}{b}|x| \leq \|x\|_{E^*} \leq c(\lambda)|x|$$

and, using Remark 2.4, it follows $M(E_\lambda) \leq \sqrt{1/\lambda}\cdot c\ell n(d_X+1)$. (Note that E_λ is the dual space to E_λ^*) . Therefore we may use Proposition 2.1 again to find for fixed $\mu < 1$ a subspace $F_\mu \subset E_\lambda^*$, $\dim F_\mu \geq \mu\lambda n$, such that

$$\frac{|x|}{c(\mu)M(E_\lambda)} \leq \|x\|_{E_\lambda^*} \leq c(\lambda)|x|$$

for any $x \in F_\mu$. It means that $d_{F_\mu} \leq c(\mu)c(\lambda)M(E_\lambda) \leq c(\mu)c(\lambda)\sqrt{1/\lambda}\cdot c\ell n(d_X+1)$. So, we have found a space F_μ such that its distance to ℓ_2^k ($k = \dim F_\mu$) is estimated by ℓn of the distance of X to ℓ_2^n . So strong majoration gives a possibility to iterate this procedure with a suitable choice of $\lambda(=\lambda_i's)$ and $\mu(=\mu_i's)$. Then we obtain the following theorem (see the details in $[M_2]$).

<u>Theorem. For any $\theta < 1$ there exists</u> $f(\theta) > 0$ <u>such that every n-dimensional normed space X contains a subspace $E \subset X$ and the dual E^* contains a subspace $F \subset E^*$ which satisfies</u>
i) $\dim F \geq \theta n$ <u>and</u> ii) $d_F \leq f(\theta)$.

4. Our next corollary we obtain using the same inner product as in Section 3. By Proposition 2.1 there is a subspace $E_\lambda \subset X$ with $|x|/c(\lambda) \leq \|x\| \leq b|x|$ for $x \in E_\lambda$. Apply to E_λ Lemma 6.1 from [F T] to find for $\mu < 1$ a subspace $E_\mu \subset E_\lambda$, $\dim E_\mu \geq \mu\lambda n$, such that

$$\|x\| \leq \frac{1}{\sqrt{1-\mu}} C_2(X)M(X)|x| \qquad \text{for } x \in E_\mu ,$$

where $C_2(X)$ is the cotype 2-constant of X . Therefore,

$$d_{E_\mu} \leq c\,\frac{c(\lambda)}{\sqrt{1-\mu}}\,C_2(X)\,\ell n(d_X+1) , \tag{4.1}$$

where c is an absolute constant. The iteration proceedure, briefly discussed in 3, can be applied again and it implies the following result

4.2 <u>Proposition. For any</u> θ , $0 < \theta < 1$, <u>there exists</u> $c(\theta) > 0$ <u>such that any n-dimensional normed space X contains a subspace $E_\theta \subset X$, $\dim E_\theta \geq \theta n$, such that</u>

$$d_{E_\theta} \leq c(\theta)C_2(X)\ell n(C_2(X)+1) .$$

4.3. Corollary. Let X be an n-dimensional normed space and let, for $2 < q < \infty$, the cotype q constant of X be at most C_q. Then for any $0 < \theta < 1$ there exists a subspace $E_\theta \subset X$ such that $k = \dim E_\theta \geq \theta n$ and $d(E_\theta, \ell_2^k) \leq c(\theta) C_q n^{1/2-1/q} \ell n n$, where $c(\theta)$ depends on θ only.

4.4. Remark. The essential part of the above Proposition is the existence of a subspace E_θ, $\dim E_\theta \geq \theta n$, such that $d_{E_\theta} \leq f(\theta; C_2(X))$ for a function f depending on $\theta < 1$ and the cotype constant $C_2(X)$ only. It was earlier observed by S.J. Dilworth [D], using the method of [M_2]. However, his function f is exponential in $C_2(X)$. (I would like also to mention that back to 1976 after [FLM] G. Pisier asked us this question).

5. The result of section 4 can be obtained using another ellipsoid (i.e. an euclidean norm $|.|$), which is also much better known, - the so called maximal volume ellipsoid. We combine Lemmas 1 and 2 from [DMT] to claim that for any space $X = (\mathbb{R}^n, \|.\|)$ there exists an Euclidean structure $(\mathbb{R}^n, |\cdot|_1)$ such that $\frac{1}{C}|x|_1 \leq \|x\| \leq |x|_1$ and $M(X^*) \leq T_2(X^*)$ where $T_2(X^*)$ is the type 2 constant of X^* (C is at most $\sqrt{n}$ but it is not used here).

Change a normalization of the norm $|\cdot|_1$ introducing $|\cdot| = |\cdot|_1 / T_2(X^*)$. Then $\|x\| \leq T_2(X^*)|x|$ and for this Euclidean norm $M(X^*) \leq 1$. Therefore, using Proposition 2.1, we have proved

5.1 Theorem. For every $0 < \lambda < 1$ any n-dimensional normed space $X = (\mathbb{R}^n, \|\cdot\|)$ contains a k-dimensional subspace E_λ for $k \geq \lambda n$ such that

$$d(E_\lambda, \ell_2^k) \leq c(\lambda) T_2(X^*) \qquad (5.2)$$

where $c(\lambda) = 2(4\pi)^{1/(1-\lambda)}$.

5.3 Remark. It is well known that $T_2(X^*) \leq C_2(X) \cdot \mathrm{Rad}_X$ and, by Pisier [P], $\mathrm{Rad}_X \leq C \ell n(d_X + 1)$. Therefore estimation (5.2) is slightly better than estimation (4.1).

5.4. Remark. Let us consider Grassmann manifold $G_{n,k}$ of k-dimensional subspaces of $\mathbb{R}^n$ and introduce on $G_{n,k}$ the normalized Haar measure associated with the Euclidean structure $(\mathbb{R}^n, |\cdot|)$. Then the proof of Proposition 2.1 and Theorem 5.1 gives also a large measure (exponentially close to the full measure) of subspaces as found in Theorem 5.1.

6. We will discuss in this section estimate (5.2). Obviously $d(E_\lambda, \ell_2^k) \geq T_2(E_\lambda^*)$, however $T_2(E_\lambda^*)$ can be significantly smaller then $T_2(X^*)$ and in this case (5.2) can be not precised. To improve this estimate we again apply some iteration pro-

ceedure.

Below we reserve letter c for any absolute constant and let $\lesssim$ denote an inequality valid after a suitable change of absolute constants involved. Also let $\simeq$ mean that both $\lesssim$ and $\gtrsim$ take place.

Fix μ close to 1, $\mu<1$. We choose μ later. We apply subsequently Theorem 5.1 many times for $\lambda=\lambda_j=\mu^{1/j}$, $j=1,2,\dots$. On step j we find a subspace $E_{\lambda_j}\subset E_{\lambda_{j-1}}$ with $k_j=\dim E_{\lambda_j}\geq \lambda_j\cdot k_{j-1}$ and $d(E_{\lambda_j},\ell_2^{k_j})\geq c(\lambda_j)T_2(E^*_{\lambda_{j-1}})$. We will write E_j instead of E_{λ_j} to simplify notations ; also $E_o\equiv X$. Note that

$$c(\lambda_j)=c^{1/(1-\lambda_j)}\simeq c^{j/\ln 1/\mu} .$$

(i) We stop our iteration if

$$T_2(E^*_{j-1})\leq c(\lambda_j)T_2(E^*_j)$$

and we have in this case

$$d_{E_j}=d(E_j,\ell_2^{k_j})\leq c(\lambda_j)^2T_2(E^*_j)\simeq c^{j/\ln 1/\mu}T_2(E^*_j) .$$

We also stop the iteration if

$$T_2(E^*_j)\leq c(\lambda_j)\simeq c^{j/\ln 1/\mu} . \tag{6.1}$$

(ii) In the opposite case

$$T_2(E^*_j)<\frac{1}{c(\lambda_j)}T_2(E^*_{j-1})<\Big(\prod_{k=1}^{j}c(\lambda_k)\Big)^{-1}T_2(X^*)\simeq \tag{6.2}$$
$$c^{-\sum_{}^{j}k/\ln 1/\mu}T_2(X^*)\simeq c^{-j^2/\ln 1/\mu}T_2(X^*) .$$

Also

$$\dim E_j\geq\prod_{k=1}^{j}\lambda_k n\simeq\mu^{\ln j}n .$$

By (6.1) and (6.2) the iteration must stop for t such that

$$t\ln c/\ln 1/\mu+t^2\ln c/\ln 1/\mu\lesssim\ln T_2(X^*) .$$

It means

$$t\lesssim\sqrt{\ln 1/\mu\cdot\ln T_2(X^*)/\ln c}\overset{\text{def}}{=}t_o .$$

Fix small $\alpha>0$. We would like to take $\mu=1-\alpha/\ln t_o$ which implies $\mu^{\ln t}\leq e^{-\alpha}$ and $\ln 1/\mu\simeq\alpha/\ln t_o$. To achieve this we define t_o from the equation

$$t_o\sqrt{\ln t_o}=\sqrt{\alpha\ln T_2(X^*)/\ln c} .$$

Therefore

$$t_o \simeq \sqrt{\frac{\ell n T_2(X^*)}{\ell n \ell n T_2(X^*)}} \cdot \frac{\alpha}{\ell n c} \; .$$

So, when the iteration has stopped we obtain a subspace $E_t \subset X$ such that

(6.3) $$\dim E_t \geq e^{-\alpha} n$$

(6.4) $$d_{E_t} \lesssim c^{t_o \ell n t_o / \alpha} \cdot T_2(E_t^*)$$

$$\simeq c^{\sqrt{\ell n T_2(X^*) \ell n \ell n T_2(X^*)/\alpha}} \cdot T_2(E_t^*) \; .$$

The formula which we have obtained still depends not only on T_2 of the dual space E_t^* but also on $T_2(X^*)$. Therefore we will continue the iteration considering as one step the above result which in itself is the result of the iteration proceedure.

Fix $\theta > 0$ and take $\alpha \equiv \theta / \ell n \ell n T_2(X^*)$. We construct a subspace E_{t_i} of a space $E_{t_{i-1}}$ as above, i.e.

(6.3') $$\dim E_{t_i} \geq e^{-\alpha} \dim E_{t_{i-1}} \geq e^{-j\alpha} n \; ,$$

(6.4') $$d_{E_{t_i}} \lesssim c^{\sqrt{\ell n T_2(E^*_{t_{i-1}}) \, \ell n \ell n T_2(E^*_{t_{i-1}})/\alpha}} \cdot T_2(E^*_{t_i}) \; .$$

If $\ell n T_2(E^*_{t_{i-1}}) \leq e \ell n T_2(E^*_{t_i})$ we stop our construction. Otherwise we proceed and build the next subspace $E_{t_{i+1}}$ from E_{t_i}. Because always $T_2 \geq 1$ the proceedure will stop after p steps where $p \lesssim \ell n \ell n T_2(X\;)$ (this is the reason for the above choice of α). So the following proposition has been proved.

6.5. Proposition. Every n-dimensional normed space X contains, for any $\theta > 0$, a subspace $F \subset X$ such that $\dim F \geq e^{-\theta} n$ and

$$d_F \leq c^{\sqrt{\ell n T_2(F^*) \ell n \ell n T_2(F^*) \ell n \ell n T_2(X^*)/\theta}} \cdot T_2(F^*) \quad ,$$

where c is an absolute constant.

6.6. Remark. Of course, we may continue the above proceedure to eliminate $\ell n \ell n T_2(X^*)$ and to exchange it on $\ell n \ell n \ell n T_2(X^*)$ and so on. We could also organize the first iterative stage more carefully to get a better result. However, our purpose was just to show all these possibilities.

REFERENCES

[D] S.J. Dilworth, The cotype constant and large Euclidean subspaces of normed spaces, Preprint.

[G] E.D. Gluskin, The diameter of the Minkowski compactum is roughly equal to n. Functional Anal. Appl., 15 (1981), 72-73.

[FLM] T. Figiel, J. Lindenstrauss, V.D. Milman, The dimension of almost spherical sections of convex bodies, Bull. Amer. Math. Soc., 82 (1976), 575-578.

[FT] T. Figiel, N. Tomĉzak-Jaegermann, Projections onto Hilbertian subspaces of Banach spaces, Israel J. Math. 33 (1979), 155-171.

[K] B.S. Kashin, Diameters of some finite dimensional sets and of some classes of smooth functions, Izv. ANSSSR, Ser. Math. 41 (1977), 334-351 (Russian).

[L] D.R. Lewis, Ellipsoids defined by Banach ideal norms, Mathematika, 26 (1979), 18-29.

[MaP] B. Maurey, G. Pisier, Séries de variables aléatoires vectorielles indépendantes et propriétés géométriques des espaces de Banach. Studia Math. 58 (1976), 45-90.

[M_1] V.D. Milman, Geometrical inequalities and Mixed Volumes in Local Theory of Banach Spaces, Astérisque, To appear.

[M_2] V.D. Milman, Almost Euclidean quotient spaces of subspaces of finite dimensional normed spaces, Proceedings Amer. Math. Soc., to appear.

[M Sch] V.D. Milman, G. Schechtman, Asymptotic Theory of Finite Dimensioanl Normed Spaces, Lecture Notes.

[P] G. Pisier, K-convexity, Proceedings of Research Workshop on Banach Spaces, The University of Iowa, 1981, 39-151.

[SzT] S. Szarek, On Kashin almost Euclidean orthogonal decomposition of ℓ_1^n, Bull. Acad. Polon Sci. 26 (1978), 691-694.

[SzT] S. Szarek, N. Tomczak-Jaegermann, On nearly Euclidean decompositions for some classes of Banach spaces, Comp. Math. 40 (1980), 367-385.

[U] P.S. Urysohn, Mean width and volume of convex bodies in an n-dimensional space, Mat. Sbornik, 31 (1924), 477-486.

RANDOM SUBSPACES OF PROPORTIONAL DIMENSION OF FINITE DIMENSIONAL NORMED SPACES: APPROACH THROUGH THE ISOPERIMETRIC INEQUALITY

V.D. Milman
Tel Aviv University and
I.H.E.S. (France)

1. INTRODUCTION AND NOTATION.

In this X will be always n-dimensional normed space $X = (\mathbb{R}^n, \|\cdot\|)$ equipped also with a euclidean norm $(\mathbb{R}^n, |\cdot|)$ and, as a consequence, with the inner product (x,y) such that $(x,x) = |x|^2$. Then the dual norm $\|\cdot\|^*$ is naturally defined by $\|x\|^* = \sup\{|(x,y)| : \|y\| \le 1\}$. Let also $d_x = d(X, \ell_2^n)$ be the Banach-Mazur distance between X and ℓ_2^n and Rad_X be the Rademacher projection of X (see [MSch] for definitions). Recall a few well known facts.

i) If X has a cotype 2 with the constant $C_2(X)$ then X contains a $(1+\varepsilon)$-isomorphic copy of ℓ_2^k where k is some proportion of n : $k \ge C(\varepsilon)n/C_2(x)^2$ (see [FLM]). In particular, it is true for $\ell_1^n (C_2(\ell_1^n) \le C$ for a numerical constant C).

ii) However, it is known [K] that for any $\lambda < 1$ and $k \simeq \lambda n$ the space ℓ_1^n contains a $C(\lambda)$-isomorphic copy of ℓ_2^k where $C(\lambda)$ depends on λ only.

The first theorem (ε-version) is proved using the isoperimetric property of the Euclidean sphere and the median of a norm. Attempts to use the same method to find large proportions of space being hilbertian have failed. The method, which was used to prove $C(\lambda)$-version (for any $\lambda < 1$), is a volume computation ([S]). Recently, the volume method was used $[M_2]$ as an intermediate step in a number of problems where large proportions $E_{\lambda n}$ of a space were important.

After my talk in the Columbia conference (cf. the preceding paper) essential progress was achieved in this topic and presented in the Séminaire de Géométrie des Espaces de Banach, Paris VII, in December 1984. It is reported in this note.

In this note we will see how the isoperimetric inequality approach from $[M_1]$ may be adapted to the problems of finding subspaces of almost full dimension with given properties (see proposition 2.5). In the same time such approach gives a principal improvement on the estimation of a function $C(\lambda)$. We also note that proposition 2.5 has a natural interpretation as a theorem about Gelfand numbers of an operator $u^{-1} : X \to \ell_2^n$. However we are not interested to develop this direction in this note.

A few more notations and facts used in the note. Let $S = S^{n-1}$ be the euclidean sphere $S^{n-1} = \{x \in X : |x|=1\}$ and μ be the normalized rotation invariant measure on S^{n-1} (i.e. $\mu(S^{n-1}) = 1$).

Let $M_X = \int_{x \in S} \| x\| \, d\mu(x)$ and the similar notation M_{X*} we use for the dual space $X^* = (\mathbb{R}^n, \| \cdot \|^*)$. The median L_X is the number such that $\mu\{x \in S : \| x\| \le L_X\} \ge 1/2$ and $\mu\{X \in S : \| x\| \ge L_X\} \ge 1/2$. It is known [FLM] that for some absolute constant $c > 0$

(1.a) $\quad |L_X - M_X| \le C$

if $\| x\| \le \sqrt{n}|x|$ (the last condition is unnecessary as T. Figiel has showed me in a private communication; however it is usually automatically satisfied).

In the proof of Proposition 2.5. we use the arguments from $[M_1]$, which are described very briefly (see also [FLM] or [MSch], chapters 2,4, for details). Note that we standardly define $(*)_\varepsilon$ being an ε-neighbourhood in the geodesic metric of a point $(*)$ on the sphere S, i.e. ε-cap; similarly, $A_p \subset S$ defines the p-neighbourhood of a set $A \subset S$. We also use $<$ and $\simeq$ (instead of $\le$ and $=$) when we are not careful to write some absolute constants involved.

2. Isoperimetric inequality technique adapted for a search of nearly euclidean sections of nearly full dimension.

We are starting from the following:

Lemma 2.1. Let $|x| = 1$ and $\| x\|^* \le \theta$. Then for any z such that $|(x,z)| \ge t$, we have $||z|| \ge t/\Theta$.

Proof. $\| z\| \cdot \theta \ge \| z\| \cdot \| x\|^* \ge |(x,z)| \ge t$ $\square$

Corollary 2.2. Let $\frac{1}{a}|x| \le \|x\| \le b|x|$ and $L_{X*} \le 1$. Then there exists $A \subset S^{n-1}$, $\mu(A) \ge 1/2$, such that for any $x \in A_{\Pi/2-\phi}$ we have

(2.a) $\|x\| \ge \sin\phi \gtrsim \phi$.

(Take $A = \{x \in S: \|x\|^* \le 1\}$ and use lemma 2.1).

Note also that condition $M_{X*} \le 1$ would imply $\|x\| \ge \phi/C$ instead of (2.a) for an absolute constant C (use (1.a)).

Remark 2.3. By the isoperimetric property of the sphere S^{n-1} we have

$$\mu(A_{\pi/2-\varepsilon}) \ge \mu((*)_{\pi-\varepsilon}) = \frac{\int_0^{\pi/2-\varepsilon}\cos^{n-2}\theta d\theta + \int_0^{\pi/2}\cos^{n-2}\theta\, d\theta}{2I_{n-2}} =$$

$$= 1 - \frac{1}{2I_n}\int_0^{\varepsilon}\sin^{n-2}\theta d\theta \ge 1 - c\sqrt{n}\,\sin^{n-2}\varepsilon,$$

for an absolute constant $c > 0$.

Remark 2.4. Let $(\pm *)$ be a pair of antipodal points on S^k. Then, for $\kappa = 1/k^{3/2}$,

$$\alpha = \mu((\pm *)_\delta) = \frac{1}{I_k}\int_0^{\delta}\sin^{k-1}t\, dt \ge \frac{1}{I_k}\int_{\delta(1-\kappa)}^{\delta}\sin^{k-1}t\, dt \gtrsim$$

$$\gtrsim \sqrt{k}\,\kappa\delta\,\sin^{k-1}\delta(1-\kappa) > \frac{1}{k}\sin^k\delta.$$

The following proposition is the main technical tool of this talk.

Proposition 2.5. Let $\frac{1}{a}|x| \le \|x\| \le b|x|$, and $M_{X*} \le 1$. Then for any $0 < \phi < 1$ there exists a subspace $E \subset X$, $\dim E = k \ge \lambda n$ for

$\lambda = 1 - \phi$

and for any $x \in E$

$c\,\phi|x| \le \|x\| \le b|x|$;

c above is a numerical constant.

Remark. We have always a bound on a in the Proposition $a \lesssim \sqrt{n}$ because $\|x\|^* \lesssim \sqrt{n}\, M_{X*}$. We use some bound on a when we ignore in the proof below the term $\frac{\log n}{n}$ as insignificant. After that moment we cannot take ϕ too close to zero and λ too close to 1. However the statement of the Proposition is automaticaly true for $\phi \le c/n$ because $a \le \sqrt{n}$.

Proof. Take $A \subset S^{n-1}$ from Corollary 2.2 and an integer $k = \lambda n$ for fixed $\frac{1}{2} < \lambda < 1$. Note that

$$(A_{\pi/2-\varepsilon})_{\varepsilon-\phi} = A_{\pi/2-\phi}$$

and apply Remark 2.3 to estimate $(A_{\pi/2-\varepsilon})$. By a standard integral geometry argument we may find a $(k+1)$-dimensional subspace $E_{k+1} \subset \mathbb{R}^n$ such that for $S^k = S(E)$

$$\beta = \mu(A_{\pi/2-\varepsilon} \cap S^k) \ge \mu(A_{\pi/2-\varepsilon}) \ge 1 - c\sqrt{n}\,\sin^n\varepsilon.$$

Clearly, if above β and α from Remark 2.4 are such that $\alpha + \beta \ge 1$ then the set $B = A_{\pi/2-\varepsilon}$ is δ-net on S^k. Take now $\delta = \varepsilon - \phi$ and use the estimation from Remark 2.4 for such δ. Then set B is $(\varepsilon-\phi)$-net on S^k if the following inequality holds

$$\text{(2.b)} \quad \sqrt{n}\,\sin^{n-2}\varepsilon < \frac{1}{k}\sin^k(\varepsilon-\phi).$$

Therefore inequality (2.b) implies that

$$S^k \subset A_{\pi/2-\phi}.$$

By Corollary 2.2 the last inclusion means that for $x \in E$

$$\frac{\phi}{c}|x| \le \|x\| \le b|x|.$$

So we have to investigate inequality (2.b) which is essentially equivalent to the following one

(2.c) $\sin\varepsilon < \sin(\varepsilon-\phi)$

or $(\sin\varepsilon)^{1/\lambda} \underset{\sim}{<} \sin\varepsilon\cos\phi - \cos\varepsilon\sin\phi$. Divide the last inequality by $\sin\varepsilon\cos\phi$ and take log. Then we have to satisfy the following inequality

$$\frac{1-\lambda}{\lambda}\log\sin\varepsilon - \log\cos\phi \underset{\sim}{<} \log(1-\mathrm{tg}\,\phi.\mathrm{ctg}\,\varepsilon).$$

Fix $\varepsilon = \varepsilon$ (say, $\varepsilon_0 = \pi/4$). For small ϕ we have $\log\cos\phi \simeq -\phi^2/2$ and $\log(1-\mathrm{tg}\phi\;\mathrm{ctg}\,\varepsilon_0) \simeq -\phi$. Therefore we see that the previous inequality is satisfied if for some absolute constant c

$$1 - \lambda \ge c\,\phi \qquad \square$$

Remark. The previous reasons (after (2.c)) were shown to me by J. Bourgain. The original proof after (2.c) was slightly longer.

We reformulate Proposition 2.5 in the following way

Theorem 2.6. Let $u : \ell_2^n \to X$ be an invertible operator. Define $M(u) = \int_S \| ux \| \, d\mu(x)$. Then for any $0 < \lambda < 1$:

1. There exists a subspace $E \subset X$ such that $\dim E = k \geq \lambda n$ and

$$\| u^{-1}|_E \| \leq c \frac{M((u^{-1})^*)}{1-\lambda}$$

for an absolute constant c.

2. (The dual version). There exists a quotient map $q : X \to X/_F$ such that $\dim X/_F \geq \lambda n$ and $\|qu\| \leq c \frac{M(u)}{1-\lambda}$

3. Iteration Lemmas.

The above Proposition 2.5 will be combined in applications with the following lemma:

Lemma 3.1. Let X be an n-dimensional space and for any subspace $E \subset X$ we may find for any $\theta < 1$ a subspace $F \subset E$ such that

$$\dim F \geq \dim E \text{ and } d_F \leq C(\theta) \log d_E.$$

Assume also $C(\theta) \geq e^2$ and $C(\theta) \geq 2 \log C(\theta^{1/2})$. Then for any $\theta < 1$ there exists a subspace $Y \subset X$ such that

$$\dim Y \geq \theta n \quad d_Y \leq e^4 C(\sqrt{\theta}) \log C(\theta^{1/4}).$$

Remark 3.2. Of course, the same Lemma as above is true if we consider subspaces of quotient spaces of X instead of subspaces only.

The proof of the Lemma is an immediate consequence of the next statement

Lemma 3.3. Let $C(\theta) \geq 2 \ln C(\theta^{1/2})$ and $C(\theta) \geq e^2$ for every $\theta \in [0,1)$ and let $1 \leq \phi(\theta) \leq n$. Let also for every $\theta < 1$

$$\phi(\theta^2) \leq C(\theta) \ln \phi(\theta). \tag{3.a}$$

Then, for any $\theta < 1$

$$\phi(\theta) \leq 2e^2 \, C(\sqrt{\theta}) \ln C(\theta^{1/4}).$$

Proof. Put $\phi(\theta) = a(\theta)\psi(\theta)$ where $a(\theta^2) = 2\, C(\theta)\ln C(\sqrt{\theta})$. Then, (3.a) implies

$$(3.b) \qquad 2 \ln C(\sqrt{\theta})\cdot\psi(\theta^2) \le \ln a(\theta) + \ln \psi(\theta).$$

Note that $\ln a(\theta) \ge 2$ and therefore if $\ln \psi(\theta) \ge 2$ we may continue the previous inequality (3.b) as $\lesssim \ln a(\theta)\cdot\ln \psi(\theta)$, which implies (since $\ln a(\theta) \le 2 \ln C(\sqrt{\theta})$)

$$\psi(\theta^2) \le \ln \psi(\theta).$$

So, if $\ln \psi(\theta^{2^j}) \ge 2$ for $1 \le j < t$ then

$$\psi(\theta^{2^t}) \le \ln\ldots\ln \psi(\theta).$$

Clearly, it may happen at most $t = t(n)$ times (for $t(n) << c\ln \ln n$ for some absolute constant c). Therefore, for any $0 < \alpha < 1$ we may take $\theta < 1$ such that $\theta_j^{2^t} = \alpha$ and $\psi(\alpha) \le e^2$. (Of course, it could happen that $\ln \psi(\theta^2) \le 2$ for some $j < t$; however (3.b) immediately implies that $\psi(\theta^{2^{j+1}}) \le 2$ and again we come to $\psi(\alpha) \le 2 < e^2$). □

We give one more "iteration" statement.

Lemma 3.4. Let a real valid function $C(\theta) \ge 1$, $0 < \theta \le 1$, satisfy the following condition: there exists $A > 1$ such that

$$(3.c) \qquad A\, C(\theta) \ge C(\sqrt{\theta})$$

for every θ, $0 < \theta < 1$. Let also for some $\alpha < 1$ and every $\theta \in (0,1)$

$$(3.d) \qquad \phi(\theta^2) \le C(\theta)\phi(\theta)^{\alpha}$$

and $\phi(\theta) \le n$. Then for every $\theta < 1$ we have

$$\phi(\theta) \le 2K(\alpha)\, C(\theta^{1/2})^{1/1-\alpha}$$

for $k(\alpha) = A^{\alpha/(1-\alpha)^2}$

Proof is similar to the proof of Lemma 3.3. Put $\phi(\theta) = a(\theta)\psi(\theta)$ for $a(\theta^2) = K\, C(\theta)^{1/1-\alpha}$. Then, from (3.d),

$$K\, C(\theta)^{\alpha/(1-\alpha)} \psi(\theta^2) \leq K^{\alpha} C(\theta^{1/2})^{\alpha/(1-\alpha)} \psi(\theta)^{\alpha}$$

Condition (3.c) implies

$$\psi(\theta^2) \leq \psi(\theta)^{\alpha}.$$

Iterating this inequality we obtain $\psi(\theta^{2^t}) \leq \psi(\theta)^{\alpha^t}$. Take t such that at the first time $\psi(\theta)^{\alpha^t} \leq 2$. Then

$$\ln \ln n \simeq t \ln 1/\alpha \quad \text{and} \quad t \simeq \ln \ln n / \ln 1/\alpha .$$

So, starting from $\theta_0 < 1$, we may take $\theta < 1$ such that $\theta^{2^t} \simeq \theta_0$. It means that $\psi(\theta_0) \leq 2$ and it ends the proof. □

4. Applications.

Theorem 4.1. Let an n-dimensional normed space X have a cotype 2-constant $C_2(X)$. For every $\lambda < 1$ there exists a subspace $E \subset X$ such that $\dim E = k \leq \lambda n$ and

$$d = d(E, \ell_2^k) \leq c \, \frac{C_2(X)}{1-\lambda} \cdot \ln \frac{C_2(X)}{1-\lambda} ,$$

where c is a numerical constant.

In another formulation: for every $d \geq 2$ there exists a subspace $E \subset X$ such that

$$d(E, \ell_2^{\dim E}) \leq d$$

and $\dim E \geq (1-\kappa) n$ for some κ

$$\kappa \leq c \, \frac{C_2(X)}{d} \log d$$

where c is an numerical constant.

Remark. A new element in the previous theorem with respect to Dilworth [D] or [M_3] is an estimation on $\kappa(d)$ or, similarly, dependence on λ in the above estimate on d.

Proof. Use Proposition 2.5 with the maximal volume ellipsoid. After a normalization $M_{X^*} = 1$, we have for this euclidean norm on X (see [DMT], Lemma 1) for any $x \in X$

$$\frac{1}{a}|x| \leq \|x\| \leq T_2(X^*)\ |x|.$$

By Proposition 2.5 we will find a subspace E of X with $\dim E = k \geq \lambda n$ for

$$\lambda = 1 - \phi$$

and such that for any $x \in E$

$$c\phi|x| \leq \|x\| \leq T_2(X^*)\ |x|.$$

Therefore $d(E,\ell_2^k) \lesssim T_2(X^*)/\phi \leq C_2(X)\,\mathrm{Rad}_X/\phi \leq \frac{1}{\phi} C_2(X)\,\ell n\, d_X$. From the formula for $\lambda = \lambda(\phi)$ we have

$$\phi \simeq (1-\lambda)$$

and therefore

$$d_E \lesssim \frac{C_2(X)}{1-\lambda}\ .\ \log d_X.$$

To finish the proof of Theorem 4.1. it remains to apply the iteration Lemma 3.1.

Theorem 4.2. Let X be any n-dimensional normed space. For any $\lambda < 1$ there exists a quotient Y of a subspace of X such that $\dim Y \geq \lambda n$ and

$$d = d(Y,\ell_2^{\dim Y}) \leq c(1-\lambda)^{-2}|\ell n(1-\lambda)|$$

for a numerical constant c.

In a different formulation : for every $d \geq 2$ there exists a subspace $E \subset X$ and a subspace $F \subset E^*$ with $\dim F = (1-\kappa)n$ for which $d_F \leq d$ and

$$k \leq c\sqrt{\left(\frac{\log d}{d}\right)}$$

(c is a numerical constant).

Remark. A new element in the above theorem with respect to $[M_2]$ is dependence on λ in the estimation on d_Y. The previously known estimation was an exponential function on $(1-\lambda)^{-1}$.

Proof. We equip X with a euclidean structure $|\cdot|$ such that $M_X \cdot M_{X^*} \leq c \, \mathrm{Rad}_X$ (for an absolute constant c). The existence of such a structure follows from [L] and [FT]. We normalize this euclidean norm such that $M_{X^*} = 1$ and $M_X \leq c \, \mathrm{Rad}_X$. Let

$$\frac{1}{a} |x| \leq \| x\| \leq b|x|$$

for $x \in X$. By Proposition 2.5 there exists a subspace E, $\dim E \geq \lambda_1 n$, such that for any $x \in E$

$$\phi|x| \leq \| x\| \leq b|x|$$

and $\lambda_1 = 1 - \phi$. Also $M_E \leq N\left(\frac{1}{\lambda_1}\right) M_X \lesssim c \, \mathrm{Rad}_X$. Change the normalization of $|\cdot|$ in E in a such way that $M_E \leq 1$ (for $|\cdot|_1 = |\cdot| \cdot c \, \mathrm{Rad}_X$). Then for $x \in E^*$

$$\frac{1}{b_1} |x|_1 \leq \| x\|^* \leq c \frac{\mathrm{Rad}_X}{\phi} |x|_1 .$$

Again use Proposition 2.5. with the same ϕ as before. Then we find a subspace $F \subset E^*$ with $\dim F \geq \lambda_1 \dim E \geq \lambda_1^2 n$ such that

$$\lambda = \lambda_1^2 = (1-\phi)^2 \simeq 1 - c_1\phi$$

and for $x \in F$

$$\phi|x| \leq \| x\|^* \leq c \frac{\mathrm{Rad}_X}{\phi} |x| .$$

This means

$$d_F \leq c \, \mathrm{Rad}_X/\phi^2 \leq \frac{c}{\phi^2} \log d_X .$$

Let $\kappa = 1 - \lambda \simeq \phi$. Then

$$d_F \leq \frac{\log d_X}{\kappa^2} .$$

It remains to use the iteration Lemma 3.1. with Remark 3.2.

REFERENCES.

[D] S.J. Dilworth, The cotype constant and large Euclidean subspaces of normed spaces, Preprint.

[DMT] W.J. Davis, V.D. Milman, N. Tomczak-Jaegermann, The distance between certain n-dimensional spaces, Israel J. Math., 39 (1981), 1-15.

[FLM] T. Figiel, J. Lindenstrauss, V.D. Milman, The dimension of almost spherical sections of convex bodies, Acta Math. 139 (1977), 53-94.

[FT] T. Figiel, N. Tomczak-Jeagermann, Projections onto Hilbertian subspaces of Banach spaces, Israel J. Math. 33 (1979), 155-171.

[K] B.S. Kashin, Diameters of some finite dimensional sets and of some classes of smooth functions, IZV. ANSSSR, Ser. Math. 41 (1977), 334-351.

[L] D.R. Lewis, Ellipsoids defined by Banach ideal norms, Mathematika, 26 (1979), 18-29.

[M_1] V.D. Milman, New proof of the theorem of Dvoretzky on sections of convex bodies, Funct. Anal. Appl. 5(1971), 28-37 (Russian); English translation:

[M_2] V.D. Milman, Almost Euclidean quotient spaces of subspaces of finite dimensional normed space, Proceedings Amer. Math. Soc., to appear. (1985)

[M_3] V.D. Milman, Volume Approach and Iteration Procedures in Local Theory of Normed Spaces, this volume.

[MSch] V.D. Milman, G. Schechtman, Asymptotic Theory of Finite Dimensional Normed Spaces, Springe Lecture Notes, to appear.

[S] S. Szarek, On Kashin almost Euclidean orthogonal decomposition of ℓ_1^n, Bull. Acad. Polon. Sci. 26 (1978), 691-694.

FACTORING OPERATORS THROUGH HEREDITARILY-ℓ^p SPACES

Richard D. Neidinger
Department of Mathematics
Davidson College
Davidson, N.C. 28036

Introduction

When one set almost absorbs another (defined below), some properties of the former set are necessary in the latter. Some properties are also "preserved" by any Tauberian operator ([KW] and below). Section 1 shows how these general principles explain some of the results in "Factoring weakly compact operators", [DFJP]. A space is called <u>hereditarily-ℓ^p</u> provided every infinite-dimensional closed subspace contains an isomorphic copy of ℓ^p. In Section 2, we identify a class of operators, called thin operators, which factor through hereditarily-ℓ^p spaces for every $1 \leq p < \infty$. This uses the construction of [DFJP] and the concept of one set almost absorbing another. Any compact operator is thin and any thin operator is strictly singular but neither converse is true.

Throughout this paper, the letters X, Y, and Z denote Banach spaces. The word <u>operator</u> means bounded linear operator and <u>subspace</u> means closed linear subspace. Let $B_X = \{x \in X: \|x\| \leq 1\}$. For $W \subset X$, let [W] denote the closed linear span of W.

<u>Definition</u>. Let W and V be subsets of a Banach space X. We say W <u>absorbs</u> V if there exists $t > 0$ such that $V \subset tW$. We say W <u>almost absorb</u> V if for every $\varepsilon > 0$, there exists $t > 0$ such that $V \subset tW + \varepsilon B_X$.

Note that if W almost absorbs V, then $V \subset [W]$. Thus we could assume $X = [W]$ in the above definition.

<u>Definition</u>. The operator $T: X \to Y$ is Tauberian provided $(T^{**})^{-1}(Y) \subset X$.

An injective Tauberian operator is called simply a Tauberian injection.

Relative weak-compactness is an example of a property which is inherited by an almost absorbed set and preserved by Tauberian operators. (In any topology, we say a set is <u>relatively compact</u> if its closure is compact.) Given $W \subset X$, the construction in [DFJP] produces a space Y

---

*This is part of the author's Ph.D. dissertation prepared at the University of Texas at Austin under the supervision of Haskell P. Rosenthal.

and a Tauberian injection $J:Y \to X$ such that W almost absorbs JB_Y. Applying the general principles just mentioned, relative weak-compactness passes from W to JB_Y and from JB_Y to B_Y. Hence, a relatively weakly compact set yields a reflexive Banach space.

Given an operator $T:Z \to X$, we let $W = TB_Z$ and perform the construction of [DFJP] using a parameter p where $1 \le p \le \infty$. This produces Banach spaces Y_p. We define T to be *thin* if TB_Z does not almost absorb the ball of any infinite-dimensional subspace. If T is thin, each constructed Y_p is hereditarily-ℓ^p. Thus thin operators factor through hereditarily-ℓ^p spaces. This generalizes Figiel's result that any compact operator factors through a subspace of an ℓ^p-sum of finite-dimensional spaces, for each p [F]. In fact, the identity $I:\ell^p \to \ell^q$ for $p < q$ is a thin operator (thus factoring through hereditarily-ℓ^r spaces for all r) which does not factor through an ℓ^r-sum of finite-dimensional spaces for $r > q$.

We thank Professor Haskell P. Rosenthal for his help in the research and preparation of this material.

1. Properties preserved by almost absorbtion and Tauberian injections.

We now present the essential properties (for use in this section) of the construction of [DFJP]. The actual construction is presented at the beginning of Section 2.

Proposition 1.1. Let W *be a bounded convex symmetric non-empty subset of* X. *If the Banach space* Y *and operator* $J:Y \to X$ *are constructed as in* [DFJP] *from* W, *then:*

(a) $W \subset JB_Y$.

(b) W *almost absorbs* JB_Y.

(c) J *is a Tauberian injection*.

Let us summarize these three properties by saying Y *is Tauberian-linked to* W *by* J. We say Y *is Tauberian-linked to* W provided there exists such an operator J. The statement, "Y is Tauberian-linked to W" is a rough approximation to the statement "B_Y is isomorphic to W". Proposition 1.1 may be considered as a factorization result [DFJP].

Corollary 1.2. If $T:Z \to X$ *is any operator, then* T *factors through a Banach space* Y *which is Tauberian-linked to* TB_Z.

Proof. By Proposition 1.1, there exists a space Y which is Tauberian-linked to TB_Z by some $J:Y \to X$. Thus $TB_Z \subset JB_Y$. Define $S:Z \to Y$ by $Sz = J^{-1}(Tz)$ for all $z \in Z$. Since $SB_Z = J^{-1}(TB_Z) \subset B_Y$, $\|S\| \le 1$. Clearly $T = JS$ is the promised factorization. □

We will show that some properties are "preserved by the Tauberian-link". First, recall that a sequence (x_i) is *weak-Cauchy* in X if and only if $(f(x_i))$ converges in $\mathbb{R}$ for every $f \in X^*$. A set, $W \subset X$, is *weakly* pre-compact if and only if every sequence (x_i) in W has a weak-Cauchy subsequence. Finally, Rosenthal's Theorem yields: ℓ^1 does not embed in X if and only if B_X is weakly pre-compact [Ro]. (Y embeds in X means Y is isomorphic to a subspace of X.)

Theorem 1.3. *Let* $W, V \subset X$ *and suppose* W *almost absorbs* V.

(1) *If* W *is relatively weakly compact, then* V *is relatively weakly compact.*

(2) *If* W *is weakly pre-compact, then* V *is weakly pre-compact.*

(3) *If* W *is relatively (norm)-compact, then* V *is relatively compact.*

(4) *If* W *is separable, then* V *is separable.*

Proof. (1). For $U \subset X$, let $\tilde{U}$ denote the weak*-closure of U in X^{**}. Recall that U is relatively weakly compact if and only if U is bounded and $\tilde{U} \subset X$. Now for each $n \in \mathbb{N}$, there exists $t_n > 0$ such that $V \subset t_n W + (1/n)B_X$. Then $\tilde{V} \subset t_n \tilde{W} + (1/n)B_{X^{**}}$ which is weak*-closed since $B_{X^{**}}$ is weak*-compact. Since $\tilde{W} \subset X$, $\tilde{V} \subset \bigcap_{n=1}^{\infty} (X + (1/n)B_{X^{**}}) = X$. Since W is bounded, V is bounded. Thus V is relatively weakly compact.

(2). Let (v_i) be a sequence in V; we must exhibit a weak-Cauchy subsequence. Now for each $n \in \mathbb{N}$, there exists $t_n > 0$ such that $V \subset t_n W + (1/n)B_X$. Thus for each $i \in \mathbb{N}$, and $n \in \mathbb{N}$, there exists $w_i^n \in t_n W$ such that $\|v_i - w_i^n\| \le 1/n$. For a fixed n, $(w_i^n)_{i=1}^{\infty}$ is a sequence in the weakly pre-compact set $t_n W$. Thus there is a weak-Cauchy subsequence $(w_{i_j}^n)_{j=1}^{\infty}$. By a diagonalization argument, we may assume that $(w_{i_j}^n)_{j=1}^{\infty}$ is weak-Cauchy for every n. Then $(v_{i_j})_{j=1}^{\infty}$ must be weak-Cauchy. Indeed, let $f \in X^*$ and $\varepsilon > 0$. Choose n such that $(\|f\|/n) < \varepsilon/3$. Choose N such that for $j, k \ge N$, $|f(w_{i_j}^n) - f(w_{i_k}^n)| < \varepsilon/3$. Then for $j, k \ge N$,

$$|f(v_{i_j}) - f(v_{i_k})| \le \|f\| \cdot \|v_{i_j} - w_{i_j}^n\| + |f(w_{i_j}^n) - f(w_{i_k}^n)| + \|f\| \cdot \|w_{i_k}^n - v_{i_k}\|$$
$$\le \|f\|/n + \varepsilon/3 + \|f\|/n < \varepsilon.$$

(3). Recall that a set, $U \subset X$, is relatively compact if and only if for every $\varepsilon > 0$, there is a finite set $F \subset X$ such that $U \subset F + \varepsilon B_X$. Fix $\varepsilon > 0$. There exists $t > 0$ such that $V \subset tW + (\varepsilon/2)B_X$. Since W is relatively compact, there is a finite set $F \subset X$ such that $W \subset F + (\varepsilon/2t)B_X$. We conclude that $V \subset t(F + (\varepsilon/2t)B_X) + (\varepsilon/2)B_X = tF + \varepsilon B_X$. Thus V is relatively compact.

(4). Since W is separable, [W] is separable. Since $V \subset [W]$, V is separable. □

Theorem 1.4. *Suppose* $T: X \to Y$ *is a Tauberian injection and* U *is a* bounded subset of X.

(1) U is relatively weakly compact if and only if TU is relatively weakly compact.

(2) U is weakly pre-compact if and only if TU is weakly pre-compact.

(3) U is separable if and only if TU is separable.

Remark. The proof will show that the full strength of Tauberian injection is not needed in any one of these statements. In fact, the original paper [KW] proves that (1) holds for every bounded set if and only if T is Tauberian (possibly non-injective). We prove that (2) and (3) hold whenever T^{**} is injective.

Many other properties of bounded sets are preserved by Tauberian injections. In [N] (cf. [NR]), we show that the following properties, for a bounded set U, could analogously be added to Theorem 1.4: U is weakly closed, U is closed convex, U is closed convex and has the RNP, U is closed convex and has the KMP, and U is weakly sequentially complete.

Proof of Theorem 1.4. (1). Necessity is trivial since T is weak-to-weak continuous. Now assume TU is relatively weakly compact. Hence $\widetilde{TU} \subset Y$. Since T^{**} is weak*-to-weak* continuous, $T^{**}(\tilde{U}) \subset \widetilde{TU} \subset Y$. By definition of Tauberian, $\tilde{U} \subset X$ and thus U is relatively weakly compact.

(2). We use the fact that T^{**} is injective or, equivalently, $T^*(Y^*)$ is dense in X^*. We prove the stronger statement that T preserves weak-Cauchy sequences. A sequence $(u_n) \subset U$ is defined to be weak-Cauchy if $(f(u_n))$ converges for every $f \in X^*$. This holds if and only if $(T^*g(u_n)) = (g(Tu_n))$ converges for every $g \in Y^*$, by hypothesis on T. The latter statement holds if and only if (Tu_n) is weak-Cauchy.

(3). The continuous image of a separable set is separable. Now assume TU is separable. We assume without loss of generality that U is convex (since U is separable if and only if co(U) is separable). Let D be a countable subset of U with TD dense in TU. Now T^{**} is an injective weak*-to-weak* continuous map on the weak*-compact set $B_{X^{**}}$. Hence $T^{**}|_{B_{X^{**}}}$ is a weak*-to-weak* homeomorphism. Thus $T|_{B_X}$ is a weak-to-weak homeomorphism. We conclude that D is weakly dense in U. Let D' be the rational coefficient convex hull of D. Then D' is a countable dense subset of U. □

Theorems 1.3 and 1.4 combine to yield Corollary 1.5. Recall that Corollary 1.5 applies when Y is constructed as in [DFJP] from the set $W \subset X$.

Corollary 1.5. Suppose Y is Tauberian-linked to a set $W \subset X$.

(1) Y is reflexive if and only if W is relatively weakly compact.

(2) ℓ^1 does not embed in Y if and only if W is weakly pre-compact.

(3) Y is separable if and only if W is separable.

Proof. We prove only (1) since (2) and (3) follow the same pattern. There exists a Tauberian injection $J:Y \to X$ such that $W \subset JB_Y$ and W almost absorbs JB_Y. Suppose W is relatively weakly compact. Then JB_Y is relatively weakly compact by Theorem 1.3(1). Hence B_Y is weakly compact by Theorem 1.4(1). Thus Y is reflexive. Conversely, suppose Y is reflexive so that B_Y is weakly compact. Then JB_Y is relatively weakly compact by Theorem 1.4(1). Since $W \subset JB_Y$, W is relatively weakly compact. □

Clearly, Corollaries 1.2 and 1.5(1) together imply the following result from [DFJP].

Corollary 1.6. Any weakly compact operator factors through a reflexive space.

Let us consider properties other than those in Corollary 1.5. It has been mentioned that the RNP and KMP are preserved by Tauberian injections. However, there exists a set W with the RNP which almost absorbs a set V which fails the KMP. In fact, if Y is constructed as in [DFJP] from the RNP-set W, then L^1 (which fails the KMP) embeds in Y. This counterexample is due to H. P. Rosenthal [N, p.112]. W. Schachermayer made a related observation in [S].

It was also mentioned that weak-sequential completeness is preserved by Tauberian injections, as is relative weak-sequential completeness. The following question remains open. If W is relatively weakly sequentially complete and W almost absorbs V, is V relatively weakly sequentially complete? If this is true, then this property will be "preserved by a Tauberian-link" as in Corollary 1.5.

Although some properties (e.g., RNP) are not inherited by an almost absorbed set, the Tauberian injection can still yield a partial result. Suppose Y is Tauberian-linked to W. Then there exists a Tauberian injection $J:Y \to [W]$. Many space properties are inherited through any Tauberian injection, i.e., if [W] has property P, then Y has P. Specifically, this is true for the RNP, the KMP, quasi-reflexivity, somewhat reflexivity, weak-sequential completeness, the property that c_0 does not embed in Y, and the property that Y* is separable (of course, also reflexivity, the property that ℓ^1 does not embed in Y, and separability) [N, p.70-94,110]. This type of result does not hold for every isomorphic property. Indeed, there exists a Tauberian injection $J:Y \to c_0$ such that Y fails property (u) even though c_0 satisfies property (u) [N, p.130].

2. Factoring operators through hereditarily-ℓ^p spaces.

Definition. Let W be a bounded convex symmetric non-empty subset of X and $1 \le p \le \infty$. We say Y_p and J_p are constructed as in [DFJP] from W if Y_p and J_p are defined as follows. For n=1,2,..., define $U_n = 2^n W + 2^{-n} B_X$ an absorbing bounded convex symmetric set. Let $|||\cdot|||_n$

be the Minkowski functional or gauge of U_n, i.e., the unique norm such that $\{x: |||x|||_n < 1\} \subset U_n \subset \{x: |||x|||_n \le 1\}$. Note that $|||\cdot|||_n$ is equivalent to $\|\cdot\|$, the original norm on X. Indeed let X_n denote X under the norm $|||\cdot|||_n$ and let $M = \sup\{\|w\| : w \in W\}$. Then $2^{-n}B_X \subset B_{X_n} \subset (2^n M + 2^{-n})B_X$. In particular, $\|x\| \le (2M + \frac{1}{2})|||x|||_1$. If $p < \infty$, let $D = (\sum_{n=1}^{\infty} \oplus X_n)_p$, the ℓ^p-sum. If $p = \infty$, let $D = (\sum_{n=1}^{\infty} \oplus X_n)_0$, the c_0-sum. Define Y_p to be the diagonal subspace of D, i.e., $Y_p = \{(x_1, x_2, \cdots) \in D: x_1 = x_2 = x_3 = \cdots\}$, or $Y_p = \{(x,x,\cdots) : x \in X,\ (|||x|||_n)_{n=1}^{\infty} \in \ell^p \text{ (or } c_0 \text{ if } p = \infty)\}$.

Observe that Y_p is closed in D and is therefore a Banach space. Define $J_p: Y_p \to X$ by $J_p(y) = x$ for $y = (x,x,\cdots) \in Y_p$. Clearly J_p is a one-to-one linear map. Also, if $J_p y = x$, then $\|x\| \le (2M + \frac{1}{2})|||x|||_1 \le (2M + \frac{1}{2})\|y\|$. Thus, $\|J_p\| \le 2M + \frac{1}{2}$.

We say Y_p and J_p <u>are constructed as in [DFJP] from an operator</u> $T: Z \to X$ if Y_p and J_p are constructed as in [DFJP] from TB_Z.

Note that for $p < \infty$, $J_p B_{Y_p} = \{x \in X: \sum_{n=1}^{\infty} |||x|||_n^p \le 1\}$.

Also, $J_\infty B_{Y_\infty} = \{x \in X: |||x|||_n \to 0 \text{ and } \sup_n |||x|||_n \le 1\}$.

Thus if $p < q$, then $J_p B_{Y_p} \subset J_q B_{Y_q}$.

Actually, only the case $p = 2$ is explicitly used in [DFJP]. One may assume $p = 2$ in the construction specified in Proposition 1.1. We now restate Proposition 1.1 to include the parameter.

<u>Proposition 2.1</u>. <u>Let</u> W <u>be a bounded convex symmetric non-empty subset of</u> X <u>and</u> $1 \le p \le \infty$. <u>If the Banach space</u> Y_p <u>and injective operator</u> J_p <u>are constructed as in [DFJP] from</u> W, <u>then</u>

(a) $W \subset J_p B_{Y_p}$

(b) W <u>almost absorbs</u> $J_p B_{Y_p}$

(c) <u>If</u> $1 < p < \infty$, J_p <u>is Tauberian</u>.

<u>Remark</u>. It is also true that J_1 is a semi-embedding ($J_1 B_{Y_1}$ is closed) and J_∞^{**} is injective.

In [DFJP], they prove that $(J_2^{**})^{-1}(X) \subset Y_2$, i.e., J_2 is Tauberian. We omit the proof of Proposition 2.1(c) (cf. [N, p.101-103]). We focus instead, on (a) and (b).

<u>Proof of Proposition 2.1 (a) and (b)</u>. Fix $1 \le p \le \infty$ and suppress

p in the notation.

To establish (a), let $w \in W$. For any $n \in \mathbb{N}$, $W \subset 2^{-n}\{x: |||x|||_n \le 1\}$. Thus $|||w|||_n \le 2^{-n}$. If $y = (w, w, \cdots)$, then $J(y) = w$ and $\|y\| = \|(|||w|||)\|_p \le \sum_{n=1}^{\infty} |||w|||_n \le 1$.

To establish (b), let $x \in JB_Y$, i.e., $\|(|||x|||_n)\|_p \le 1$. Thus, for all n, $|||x|||_n \le 1$ and since $|||\cdot|||_n$ is the gauge of U_n, $x \in \overline{U}_n = \overline{2^n W + 2^{-n}B_X}$. Thus for $\varepsilon > 0$, choose n so that $2^{-n} < \varepsilon$, then $JB_Y \subset 2^n W + \varepsilon B_X$. (Moreover, if $1 \le p < \infty$, each $|||x|||_n < 1$, so that $x \in U_n$. Thus, $JB_Y \subset 2^n W + 2^{-n} B_X$ for all n.) □

If Y_p and $J_p: Y_p \to X$ are constructed as in [DFJP] from $T: Z \to X$, then TB_Z almost absorbs $J_p B_{Y_p}$ and $TB_Z \subset J_p B_{Y_p}$. This latter containment implies that T factors through Y_p using J_p. Since Y_p is a subspace of $(\sum_{n=1}^{\infty} \oplus X_n)_p$, Y_p may be expected to have ℓ^p subspaces. When is Y_p hereditarily-ℓ^p? Specifically, what property on $T: Z \to X$ will imply that every Y_p, constructed as in [DFJP] from T, is hereditarily-ℓ^p? We identify such a property. This property is between compactness and strict singularity.

Recall that $T: Z \to X$ is <u>strictly singular</u> if there does not exist an infinite-dimensional subspace $\hat{Z}$ of Z such that $T|_{\hat{Z}}$ is an isomorphism. Suppose T is injective. Then, by the open mapping theorem, T is strictly singular if and only if there does not exist an infinite-dimensional subspace $\hat{X}$ of X such that TB_Z absorbs $B_{\hat{X}}$.

<u>Definition</u>. An operator $T: Z \to X$ is <u>thin</u> if there does not exist an infinite-dimensional subspace $\hat{X}$ of X such that TB_Z almost absorbs $B_{\hat{X}}$.

Of course, $T: Z \to X$ is compact if and only if TB_Z is relatively compact. We present the main theorem of Section 2.

<u>Theorem 2.2. Let</u> $T: Z \to X$ <u>be an operator. Consider the following conditions on</u> T:

(a) T <u>is compact</u>.

(b) T <u>is thin</u>.

(c) <u>For every</u> $1 \le p \le \infty$, <u>if</u> Y_p <u>is constructed as in</u> [DFJP] <u>from</u> T, <u>then</u> Y_p <u>is hereditarily</u>-ℓ^p <u>or hereditarily</u>-c_0 <u>in the case</u> $p = \infty$.

(d) <u>For every</u> $1 \le p \le \infty$, T <u>factors through a hereditarily</u>-ℓ^p (c_0 if $p = \infty$) <u>space</u>.

(e) T <u>is strictly singular</u>.

<u>Then</u> (a) ⇒ (b) ⇒ (c) ⇒ (d) ⇒ (e).

For the purpose of discussion, we will summarize conditions (c) and (d) as follows. We say T yields hereditarily-ℓ^p Y_p's if and only if T satisfies condition (c) of Theorem 2.2. We say T factors through hereditarily-ℓ^p spaces if and only if T satisfies condition (d).

Before proving Theorem 2.2, we make a few comments. Compact, thin, and strictly singular are distinct concepts. We will show that the identity operator $T:\ell^p \to \ell^q$, where $1 < p < q < \infty$, is an example of a thin operator which is not compact. We will also show that the identity operator $T:L^\infty \to L^1$ is strictly singular but not thin. In fact, $T:L^\infty \to L^1$ does not yield hereditarily-ℓ^p Y_p's. However, the veracity of other reverse implications remains undecided. Specifically, the property that T factors through hereditarily-ℓ^p spaces may be equivalent to thin or strictly singular or all five conditions may be distinct.

Theorem 2.2 generalizes a result of Figiel which followed the work of Johnson [J]. For $1 \le p < \infty$, let $F_p = (\sum_{n=1}^{\infty} \oplus \ell_n^\infty)_p$, the ℓ^p-sum of n-dimensional spaces, which is clearly hereditarily-ℓ^p. Let $F_\infty = c_0$. Figiel showed that any compact operator factors through a subspace of F_p for every $1 \le p \le \infty$ [F]. We show that any thin operator T factors through a hereditarily-ℓ^p (c_0 if $p = \infty$) space Y_p for every $1 \le p \le \infty$. We point out three differences between these results. First, the class of thin operators encompasses the class of compact operators. Second, Y_p depends on T and F_p does not. Third, Y_p may not be isomorphic to an ℓ^p-sum of finite-dimensional spaces and, thus, is a different "kind" of hereditarily-ℓ^p space. The referee's ideas, concerning this latter point, are included in the following paragraph.

Let $1 < p < q < r < \infty$. Any operator from a subspace of an ℓ^r-sum of finite-dimensional spaces to ℓ^q must be compact. (This follows from the same reasoning used for an operator from ℓ^r to ℓ^q [LT,p.76].) Since the identity map $T:\ell^p \to \ell^q$ is not compact, T does not factor through any subspace of an ℓ^r-sum of finite-dimensional spaces (including F_r). Since T is thin, T does factor through Y_r which is hereditarily-ℓ^r. Thus Y_r is not isomorphic to any subspace of an ℓ^r-sum of finite-dimensional spaces. In general T is compact implies T factors through a subspace of F_p for every $1 \le p \le \infty$, but the converse is false. Indeed, the identity map $T:\ell^1 \to c_0$ is not compact (T is thin) but trivially factors through ℓ^p (a subspace of F_p) for every $1 < p < \infty$. We leave open the question: if T factors through a subspace of F_p for every $1 \le p \le \infty$, is T thin?

Proof of Theorem 2.2. (a) $\Rightarrow$ (b). Assume T is compact, i.e., TB_Z is relatively compact. Suppose T is not thin. Then there exists an infinite-dimensional subspace $\hat{X}$ of X such that TB_Z almost absorbs $B_{\hat{X}}$. By Theorem 1.3 (1), $B_{\hat{X}}$ is compact. This contradicts $\hat{X}$ is infinite-dimensional.

(b) $\Rightarrow$ (c). Fix $1 \le p < \infty$ and suppress p in the notation. (The case $p = \infty$ is proved similarly.) Let Y and J be constructed as in [DFJP] from T. Suppose J is not strictly singular, i.e., there exists an infinite-dimensional subspace $\hat{Y}$ of Y such that $J|_{\hat{Y}}$ is an isomorphism. Let $\hat{X} = J(\hat{Y})$, an infinite-dimensional subspace of X. Then $JB_{\hat{Y}}$ absorbs $B_{\hat{X}}$. By Proposition 2.1, TB_Z almost absorbs JB_Y. Hence, TB_Z almost absorbs $B_{\hat{X}}$. This means that T is not thin, which contradicts the hypothesis. We conclude that J is strictly singular.

Now let $\hat{Y}$ be an arbitrary infinite-dimensional subspace of Y. We must show that ℓ^p embeds in $\hat{Y}$. Now $J|_{\hat{Y}}$ is not an isomorphism. Thus for every $\delta > 0$, there exists $y = (x,x,\cdots) \in \hat{Y}$ such that $\sum_{n=1}^{\infty} |||x|||_n^p = \|y\|^p \ge 1$ and yet $\|Jy\| = \|x\| \le \delta$. Recall that $|||\cdot|||_n$ is equivalent to $\|\cdot\|$ for every n (although the equivalence constant does grow as $n \to \infty$). Therefore, given any $m \in \mathbb{N}$ and $\varepsilon > 0$, we may choose δ sufficiently small so that if $\|x\| \le \delta$, then $\sum_{n=1}^{m} |||x|||_n^p < \varepsilon$. In imprecise terms let us say $y = (x,x,\cdots)$ has "mass supported on $[m,k]$" if $\|y\|^p \approx \sum_{n=m}^{k} |||x|||_n^p$. Our observations provide that for any m, there is an element $\hat{Y}$ whose mass is supported on $[m+1,\infty]$. This allows what has been called a "gliding hump argument". Choose $y_1 \in \hat{Y}$; necessarily there is a k_1 so that y_1 has mass supported on $[1,k_1]$. Choose $y_2 \in \hat{Y}$ with mass supported on $[k_1+1,k_2]$. Iterate while improving the precision on each step and one can obtain a pertubation of a disjointly supported norm-one sequence in the ℓ^p-sum. Thus (y_i) is equivalent to the usual ℓ^p basis. Therefore ℓ^p embeds in $\hat{Y}$.

(c) $\Rightarrow$ (d) is trivial since T factors through every Y_p.

(d) $\Rightarrow$ (e). Assume T factors through E_1 which is hereditarily-ℓ^p and through E_2 which is hereditarily-ℓ^q where $p \ne q$. Say $T = V_iU_i$ where $U_i: Z \to E_i$ and $V_i: E_i \to X$ for $i = 1$ and 2. Now E_1 and E_2 are totally incomparable, i.e., there is no infinite-dimensional subspace of E_1 which is isomorphic to a subspace of E_2 [LT,p.75]. Suppose that T is not strictly singular. Then there exists an infinite-dimensional subspace $\hat{Z}$ of Z such that $T|_{\hat{Z}}$ is an isomorphism. Then $U_i|_{\hat{Z}}$ is an isomorphism for $i = 1,2$. Thus E_1 has an infinite-dimensional subspace $U_1(\hat{Z})$ which is isomorphic to $U_2(\hat{Z}) \subset E_2$. From this contradiction, we conclude that T is strictly singular. □

The above proof of (b) $\Rightarrow$ (c) shows that if J_p is strictly singular, then Y_p is hereditarily-ℓ^p. For a fixed p, the converse is not always true; for example, J_2 could be an isomorphism between Y_2 and $X = \ell^2$. We claim that if Y_p is hereditarily-ℓ^p for every $1 \le p < \infty$, then J_p is strictly singular for every $1 \le p < \infty$. Indeed, consider any J_p and let $p < q < \infty$.

Then $J_p = J_q I$ where $I: Y_p \to Y_q$ is the identity. Since J_p factors through Y_p and Y_q, the argument of (d) $\Rightarrow$ (e) shows that J_p is strictly singular (assuming Y_p is hereditarily-ℓ^p and Y_q is hereditarily-ℓ^q). We conjecture that the case $p = \infty$ can be added to the above claim.

We now consider thin operators which are not necessarily compact. Since relative weak-compactness is inherited by an almost absorbed set, this yields a source of thin operators.

Proposition 2.3. *If* $T: Z \to X$ *is weakly compact and* X *has no reflexive infinite-dimensional subspace, then* T *is thin.*

Proof. Suppose TB_Z almost absorbs $B_{\hat{X}}$ where $\hat{X}$ is a subspace of X. Since TB_Z is relatively weakly compact, $B_{\hat{X}}$ is weakly compact by Theorem 1.3 (1). Hence $\hat{X}$ is reflexive and must be finite-dimensional by hypothesis. We conclude that T is thin. □

Proposition 2.3 is interesting with $X = c_0$. Many operators into c_0 are weakly compact (hence thin) but not compact (cf. [D,p.114]). If $T: Z \to \ell^1$ is any operator, then compact, weakly compact, thin, and strictly singular are all equivalent properties [M].

We also consider thin operators into ℓ^p. This requires the following instructive lemma. This lemma separates the steps involved in a standard proof of a general open mapping theorem. The proof is omitted (cf. [N,p.119] or [Ru,p.47]). Implication (1) $\Rightarrow$ (3) was formulated as the "open mapping lemma" by H. P. Rosenthal.

Lemma 2.4. *Let* $T: Z \to X$ *be an operator between Banach spaces. The following are equivalent:*

(1) *There exist* $1 > \varepsilon > 0$ *and* $t > 0$ *such that* $B_X \subset tTB_Z + \varepsilon B_X$.

(2) TB_Z *almost absorbs* B_X.

(3) TB_Z *absorbs* B_X.

(4) T *is onto.*

If W is any closed bounded convex symmetric subset of X, then there exists a Banach space Z and operator $T: Z \to X$ such that $TB_Z = W$ [BR,p.156]. Thus (1) through (3) of Lemma 2.4 are equivalent with W replacing TB_Z. As stated, Lemma 2.4 is slightly more general because TB_Z is not assumed to be closed.

Theorem 2.5. *Let* $1 \leq p < \infty$ *and* $T: Z \to \ell^p$ *be an operator. If* $(\ell^p)^*$ *does not embed in* Z^*, *then* T *is thin.*

Remarks. Theorem 2.5 holds with c_0 replacing ℓ^p. Also there is a result analogous to Theorem 2.5: for any operator $T: Z \to \ell^p$, if ℓ^p does not embed in Z, then T is strictly singular. Indeed, suppose $T|_{\hat{Z}}$ is an isomorphism for some infinite-dimensional subspace $\hat{Z}$ of Z. Since $T(\hat{Z})$ is a subspace of ℓ^p, ℓ^p embeds in $T(\hat{Z})$. Since $T(\hat{Z})$ is isomorphic to $\hat{Z}$, ℓ^p embeds in Z.

Proof of Theorem 2.5. We prove the contrapositive. Suppose TB_Z almost absorbs $B_{\hat{X}}$ for some infinite-dimensional subspace $\hat{X}$ of ℓ^p. By passing to a further subspace, we may assume that $\hat{X}$ is isomorphic to ℓ^p and there exists a projection P from ℓ^p onto $\hat{X}$ [LT, p.53]. Let $\varepsilon = 1/(2\|P\|)$. Then there is a $t > 0$ such that $B_{\hat{X}} \subset tTB_Z + \varepsilon B_{\ell^p}$. Thus $PB_{\hat{X}} \subset tPTB_Z + \varepsilon PB_{\ell^p}$. Now $PB_{\hat{X}} = B_{\hat{X}}$ and $PB_{\ell^p} \subset \|P\| B_{\hat{X}}$. Therefore $B_{\hat{X}} \subset tPTB_Z + (1/2)B_{\hat{X}}$. Consider the operator $PT: Z \to \hat{X}$. By the open mapping lemma (Lemma 2.4), PT is onto. Hence $(PT)^*: \hat{X}^* \to Z^*$ is an isomorphism and $(\ell^p)^*$ embeds in Z^*. □

For example, if $T: \ell^q \to \ell^p$ is any operator with $q \neq p$ and $1 < q < \infty$, then T is thin by Theorem 2.5. If $q > p$, then T is compact [LT, p. 76]. However, if $1 < q < p < \infty$, these thin operators are not necessarily compact.

We now focus on the difference between thin and strictly singular. Any operator onto an infinite-dimensional Banach space is not thin by Lemma 2.4. Let $T: \ell^1 \to \ell^2$ be a surjective operator. Then T is strictly singular by not thin. Let $1 \leq p \leq \infty$. If Y_p and J_p are constructed as in [DFJP] from T, then J_p is also surjective. Since J_p is always injective, Y_p is isomorphic to ℓ^2 for every p. We conclude that T is strictly singular but T does not yield hereditarily-ℓ^p Y_p's. This example exploits the non-injectivity of T.

For injective operators, the difference between thin and strictly singular seems more interesting. If T is injective, then T is not strictly singular if and only if TB_Z absorbs the ball of some infinite-dimensional subspace. Of course, T is not thin if and only if TB_Z almost absorbs the ball of some infinite-dimensional subspace. These conditions are not equivalent, even though the similar statements in Lemma 2.4 are equivalent. Indeed, consider the following example discovered by H. P. Rosenthal. Recall that $T: Z \to X$ is a semi-embedding provided T is injective and TB_Z is closed.

Proposition 2.6. Let $T: L^\infty \to L^1$ be the identity operator. Then T is a semi-embedding which is strictly singular but not thin.

Proof. First we observe that T is a semi-embedding. Let (f_n) be a sequence in TB_{L^∞} converging to f in L^1. Let $\varepsilon > 0$ and $E = \{t \in [0,1]: |f(t)| \geq 1 + \varepsilon\}$. Choose n such that $\varepsilon^2 \geq \|f_n - f\|_{L^1}$. Then $\varepsilon^2 \geq \int_E |f_n - f| \geq \varepsilon\, m(E)$ and $m(E) \leq \varepsilon$. Thus $m(\{t: |f(t)| > 1\}) = 0$ and $f \in TB_{L^\infty}$. We also note that T is weakly compact. A bounded subset A of L is relatively weakly compact if and only if A is uniformly integrable ([DU], p.76), i.e., for all $\varepsilon > 0$, there is a $\delta > 0$ such that whenever $m(E) < \delta$ and $f \in A$, then $\int_E |f| < \varepsilon$. Obviously TB_{L^∞} is uniformly integrable.

To show T is strictly singular we use the fact that T is Dunford-Pettis, i.e., T maps weakly compact sets to norm compact sets (cf. [DU], p.154 and 178). Suppose that Z is a subspace such that $T|_Z$ is an isomorphism. Of course $TB_Z \subset TB_{L^\infty}$ which is relatively weakly compact. Since $T|_Z$ is an isomorphism, B_Z is weakly compact. Since T is Dunford-Pettis, TB_Z is norm compact and again by isomorphism, B_Z is norm compact. Thus Z must be finite-dimensional and T is strictly singular. Thus TB_{L^∞} never absorbs the ball of an infinite-dimensional subspace.

We now show that TB_{L^∞} almost absorbs the ball of any reflexive subspace of L^1. Let X be a reflexive subspace of L^1 and $\varepsilon > 0$. Since B_X is weakly compact, it is uniformly integrable; hence there exists a $\delta > 0$ such that whenever $m(E) < \delta$ and $f \in B_X$, then $\int_E |f| < \varepsilon$. We claim that $B_X \subset (1/\delta)TB_{L^\infty} + \varepsilon B_{L^1}$, i.e., TB_{L^∞} almost absorbs B_X. Let $f \in B_X$. Define $E = \{t \in [0,1]: |f(t)| > 1/\delta\}$ and $E^c = [0,1] \setminus E$. Let $\tilde{f} = f\chi_{E^c}$, then $\tilde{f} \in (1/\delta)TB_{L^\infty}$ and $\int|f - \tilde{f}| = \int_E |f|$. But $1 \geq \int_E |f| > (1/\delta)m(E)$. Thus $m(E) < \delta$ and $\int |f - \tilde{f}| < \varepsilon$. □

If $T: Z \to X$ is a semi-embedding and Y_p is constructed as in [DFJP] from T, then Y_p may be identified with the Lions-Peetre interpolation space $(Z,X)_{p^{-1},p}$. The author learned this from C. Stegall and is unsure who first observed this. Let $T: L^\infty \to L^1$ be the natural injection. Let $1 < p < \infty$ and Y_p be constructed as in [DFJP] from T. Then Y_p is isomorphic to $(L^\infty, L^1)_{p^{-1},p}$ which is isomorphic to L^p [BL, p.109 & 46]. Hence, $T: L^\infty \to L^1$ does not yield hereditarily-ℓ^p Y_p's.

In [N, p.136], we present a direct example (not using the interpolation identification) of a strictly singular injection which does not yield hereditarily-ℓ^p Y_p's.

We now consider the open question: Is every strictly singular Tauberian injection necessarily thin? This question is interesting because a positive answer would imply that (b)$\Leftrightarrow$(c) in Theorem 2.2, i.e., T is thin if and only if T yields hereditarily-ℓ^p and Y_p's. Indeed, if T yields hereditarily-ℓ^p Y_p's then (as argued following the proof of Theorem 2.2) J_p is strictly singular for every $1 \leq p < \infty$. Fix $1 < p < \infty$. Since J_p is a Tauberian injection, a positive answer to the above question would imply that J_p is thin. Now $TB_Z \subset J_pB_{Y_p}$. Thus if TB_Z almost absorbs the ball of some subspace $\hat{X}$ of X, then $J_pB_{Y_p}$ almost absorbs $B_{\hat{X}}$. Assuming J_p is thin, $\hat{X}$ must be finite-dimensional. We conclude that T is thin.

REFERENCES

[BL] J. Bergh and J. Löfström, Interpolation Spaces, an Introduction, Springer-Verlag, New York, 1976.

[BR] J. Bourgain and H. P. Rosenthal, "Applications of the theory of semi-embeddings to Banach space theory," J. Functional Analysis 52 (1983), 149-188.

[D] J. Diestel, Sequences and Series in Banach Spaces, Springer-Verlag, New York, 1984.

[DFJP] W. J. Davis, T. Figiel, W. B. Johnson and A. Pelczynski, "Factoring weakly compact operators," J. Functional Analysis 17 (1974), 311-327.

[DU] J. Diestel and J. J. Uhl, Jr., Vector Measures, A.M.S. Surveys, No. 15, Amer. Math. Soc., Providence, R.I., 1977.

[F] T. Figiel, "Factorization of compact operators and applications to the approximation problem," Studia Math. 45 (1973), 191-210.

[J] W. B. Johnson, "Factoring compact operators," Israel J. Math. 9 (1971), 337-345.

[KW] N. Kalton and A. Wilansky, "Tauberian operators on Banach spaces," Proc. A.M.S. 57 (1976), 251-255.

[LT] J. Lindenstrauss and L. Tzafriri, Classical Banach Spaces I, Springer-Verlag, New York, 1977.

[M] V. D. Mil'man, "Some properties of strictly singular operators," Functional Analysis and its Applications (translated from Russian) 3 (1969), 77-78.

[N] R. D. Neidinger, "Properties of Tauberian operators on Banach spaces," Ph.D. Dissertation, University of Texas at Austin, 1984.

[NR] R. D. Neidinger and H. P. Rosenthal, "Norm-attainment of linear functionals on subspaces and characterizations of Tauberian operators," to appear in Pacific J. Math.

[Ro] H. P. Rosenthal, "Some recent discoveries in the isomorphic theory of Banach spaces," Bulletin A.M.S. 94 (1978), 803-831.

[Ru] W. Rudin, Functional Analysis, McGraw-Hill, New York,1973.

[S] W. Schachermayer, "The sum of two Radon-Nikodym-sets need not be a Radon-Nikodym-set," preprint.

THE LIE ALGEBRA OF A BANACH SPACE

Haskell Rosenthal*
Department of Mathematics
The University of Texas at Austin
Austin, Texas 78712

Introduction.

Let B be a real or complex Banach space and T be an operator (= bounded linear map) from B into itself. T is said to be skew-Hermitian if $\mathcal{R}e[f(Tb)] = 0$ for all $f \in B^*$ and $b \in B$ with $f(b) = \|f\|\|b\|$. The Lie-algebra of B, $\mathcal{A}(B)$, is defined to be the set of skew-Hermitian operators on B. If B is a complex space, we may define T to be Hermitian if iT is skew-Hermitian. Evidently for complex spaces, the study of these two classes of operators is the same, and considerable information is known about the structure of Hermitian operators (see [2] for standard material). The novelty of our approach is to focus on the structure of real Banach spaces using the Lie algebra; we obtain isometric information concerning the spaces and set up some apparently rather deep open problems.

For example, we prove the following result in Section 3: Let B be an n-dimensional real Banach space. Then B is Euclidean provided $\dim \mathcal{A}(B) > \frac{(n-1)(n-2)}{2}$ (in which case $\dim \mathcal{A}(B) = \frac{n(n-1)}{2}$). (A Banach space is said to be Euclidean if it is linearly isometric to a Hilbert space.) Moreover there exists for all n, an n-dimensional Banach space B with $\dim \mathcal{A}(B) = \frac{(n-1)(n-2)}{2}$. (In fact, we prove in [7] that $\dim B = \frac{(n-1)(n-2)}{2}$ if and only if B is a non-Euclidean rotation space, as defined below.)

Let so_n denote the Lie-algebra of $n \times n$ skew symmetric matrices. Of course so_n may be identified with $\mathcal{A}(\mathbb{R}^n)$ (where $\mathbb{R}^n$ is endowed with the usual Euclidean norm, denoted $\|\cdot\|_E$). We say that if $\|\cdot\|$ is a norm on $\mathbb{R}^n$ and $B = (\mathbb{R}^n, \|\cdot\|)$, then $\|\cdot\|_E$ and $\|\cdot\|$ are compatible if every isometry of B is an isometry of $(\mathbb{R}^n, \|\cdot\|_E)$.

---

*The research for this work was supported in part by NSF-MCS-8303534, and took place at the Weizmann Institute of Science during a Faculty Research Assignment from The University of Texas at Austin, Spring 1984.

We show in Section three that if $\|\cdot\|_E$ and $\|\cdot\|$ are compatible, then $A(B) \subset so_n$; in fact, $A(B)$ is then a Lie-subalgebra of so_n. Call the Lie-subalgebras of so_n which arise in this way, B. Lie algebras. We sketch the classical proof that given any n-dimensional Banach space B, there is a norm $\|\cdot\|$ on $\mathbb{R}^n$ with $\|\cdot\|_E$ and $\|\cdot\|$ compatible so that B is isometric to $(\mathbb{R}^n, \|\cdot\|)$; we also show that if $\|\cdot\|_1$ and $\|\cdot\|_2$ are each compatible with $\|\cdot\|_E$ and B_1 is isometric to B_2, then $A(B_1)$ is orthogonally equivalent to $A(B_2)$ (where $B_i = (\mathbb{R}^n, \|\cdot\|)_i$ for $i = 1, 2$); that is, there is an orthogonal matrix U with $U^* A(B_1) U = A(B_2)$. The general problem then arises: Classify the B. Lie algebras up to orthogonal equivalence. A complete solution to this problem ought to answer the following question, which is open as far as I know: What are the possible dimensions of the B. Lie-subalgebras of so_n? (This is the same question as: What are the possible dimensions of the group of isometries of an n-dimensional Banach space? The general problem thus also involves answering the question: What are the possible groups which arise as the connected components of the isometries of such a space?) Robbin in [5] gives an elegant intrinsic description of those Lie-subalgebras of so_n which are B. Lie-algebras. We review this description in Section 3, and also classify the one-dimensional B. Lie-subalgebras of so_n up to orthogonal equivalence. (This result is easily deduced from Robbin's discussion; however it is noteworthy that there are infinitely many such non-equivalent subalgebras of so_n for all $n \geq 4$.)

To formulate some further isometric results, we introduce the concept of a functional unconditional sum of Banach spaces. First recall that a normalized sequence $\underline{u} = (u_j)$ in a Banach space U is called a one-unconditional basis for U if $U = [u_j]$ ($[u_j]$ denotes the closed linear span of $\{u_j : j = 1, 2, \ldots\}$) and for all n and scalars $c_1, \ldots, c_n$ and $\varepsilon_1, \ldots, \varepsilon_n$ with $|\varepsilon_i| = 1$ for all i, $\|\sum_{i=1}^n c_i u_i\| = \|\sum_{i=1}^n c_i \varepsilon_i u_i\|$. (There is a similar definition for finite sequences or in fact for transfinite sequences of arbitrary cardinality. Given a finite or infinite sequence $X_1, X_2, \ldots$ of non-zero Banach spaces and a 1-unconditional basis $\underline{u} = (u_j)$ for some space U, we let $(\Sigma \oplus X_j)_{\underline{u}}$ denote the Banach space of all sequences (x_j) with $x_j \in X_j$ for all j and $\Sigma \|x_j\| u_j \in U$, under the norm $\|(x_j)\| = \|\Sigma \|x_j\| u_j\|_U$. We say that X is a functional unconditional sum of (X_j) if X is (isometric to) $(\Sigma \oplus X_j)_{\underline{u}}$ for some unconditional basic sequence $\underline{u}$. Given X and subspaces (= closed linear submanifolds) $X_1, X_2, \ldots$ of X, we say that (X_j) is a functional unconditional decomposition of X if the X_j's are all non-zero with closed linear span equal to X and for all n and $x_1, x_1', \ldots, x_n, x_n'$ with $x_i, x_i' \in X_i$ and $\|x_i\| = \|x_i'\|$ for all i, $\|\sum_{i=1}^{n} x_i\| = \|\sum_{i=1}^{n} x_i'\|$. (That is, the norm is a function of the norm of the components.) It is easily proved that X is a functional unconditional sum of (X_j) if and only if there exists a functional unconditional decomposition (Y_j) of X with Y_j isometric to X_j for all j.

We have obtained the following result, whose proof shall be presented elsewhere [7].

<u>Theorem</u>. <u>Let</u> X <u>be a real separable Banach space</u>. <u>The following are equivalent</u>:

(a) X <u>is a functional unconditional sum of non-one-dimensional separable Hilbert spaces</u>.

(b) X <u>is the closed linear span of the ranges of the rank</u>-2 <u>members of its Lie algebra</u>.

(The theorem may be viewed as the real analogue of a theorem of Kalton and Wood [4].) Let us call the class of such Banach spaces FHS (for Functional Hilbertian Sums). We have obtained a complete isometric classification of all such spaces (as well as the non-separable analogues); it turns out they have isometrically unique 1-unconditional bases (up to permutation of course). Note that if B is a separable complex Banach space with a one-unconditional basis, then if $B_{\mathbb{R}}$ is B regarded as a real Banach space, $B_{\mathbb{R}}$ is a functional unconditional sum of 2-dimensional Hilbert spaces. Thus a space X in FHS admits a complex structure (i.e., is isometric to $B_{\mathbb{R}}$ for some complex B) if and only if it is a functional unconditional sum of even-dimensional Hilbert spaces.

We have obtained the solution to the B. Lie algebra classification problem for spaces of dimension up to four. We formulate the solution in terms of the isometric structure of the Banach spaces themselves. We first need the following concept: say that a real Banach space X is a <u>rotation space</u> if X is a functional unconditional sum of a Hilbert space H and $\mathbb{R}$, with $\dim H \geq 2$. It follows from our results in Section 3 that if X is a non-Euclidean n-dimensional rotation space and $A(X)$ is a B. Lie algebra, then $A(X)$ is orthogonally equivalent to so_{n-1} regarded as canonically embedded in so_n.

The classification of these low dimensional spaces goes as follows: Let B be a real Banach space with $A(B) \neq 0$. Then evidently $\dim B > 1$. If $\dim B = 2$, $\dim A(B) = 1$ and B is Euclidean. If B is 3-dimensional, either $\dim A(B) = 1$ and then B is a non-Euclidean rotation space or $\dim A(B) = 3$ and B is Euclidean.

(These results are quite easy to prove and are established in Section 3.) Now suppose B is 4-dimensional. There are infinitely many non-equivalent possibilities if $\dim A(B) = 1$. As noted above, if $\dim A(B) \geq 4$, B is Euclidean and then $\dim A(B) = 6$. The remaining possibilities are proved elsewhere [7]. They are as follows: if $\dim A(B) = 3$, B is a non-Euclidean rotation space. Finally, there is exactly one 2-dimensional B. Lie subalgebra of so_4. In fact, if $\dim A(B) = 2$, there is a 2-dimensional complex non-Euclidean Banach space X with a one-unconditional (normalized) basis (u_1, u_2) with B isometric to $X_{\mathbb{R}}$, and then $A(B)$ is orthogonally equivalent to $so_2 \oplus so_2$.

We wish finally to sketch the organization of the work below. In Section one, we review preliminary work concerning the Lie algebra of a Banach space, numerical ranges of operators, etc. We also present a simple proof of the known fact that for an operator T on a complex Banach space, the norm of T is bounded by e times the spectral radius of T. In Section two, we establish several equivalent conditions for a real Banach space to admit a complex structure (isometrically). One of these equivalences seems to lie rather below the surface and is proved by using the complexification of a real Banach space (this is the only place in the paper where we use complex methods to prove a theorem about real Banach spaces). In the third section, we obtain the Lie-algebra-characterization of Euclidean spaces mentioned above. We also classify all one-dimensional B-Lie algebras and give the structure of a general element of the Lie algebra of a finite-dimensional space.

Most of this work was done during a stay at the Weizmann Institute of Science. I wish to thank the members of the Theoretical Mathematics Department for their warm hospitality and scientific support during this visit.

§1. Preliminaries

We present here some basic equivalences and permanence properties of the Lie algebra of a Banach space. Of course this does involve results concerning the numerical range of an operator. We first recall some of these notions; proofs of some of the elementary properties are included for completeness and ease in readability. (For example we give a simple proof of the equivalence of the norm and numerical radius on complex Banach spaces.)

For B a (real or complex) Banach space, we let

$$\Pi(B) = \{(b^*, b) : b^* \in B^*, \ b \in B \text{ and } \|b^*\| = \|b\| = b^*(b) = 1\} .$$

(When B is understood, we set $\Pi(B) = \Pi$; also for $f \in B^*$ and $b \in B$, we sometimes set $< f , b > = f(b)$.) $\mathcal{L}(B)$ denotes the space of all bounded linear operators on B. (We will usually simply refer to the members of $\mathcal{L}(B)$ as operators.)

We now list several definitions and notations to be used in the sequel.

Definitions. Fix B and $T \in \mathcal{L}(B)$.

1. The Spatial Numerical Range of T, denoted $SNR(T)$, is defined to be $\{< b^* , Tb > : (b^* , b) \in \Pi(B)\}$.
2. The Numerical Range of T, denoted $NR(T)$, is defined as $\overline{co}\, SNR(T)$.
3. The numerical radius of T, denoted $nr(T)$, is defined by $nr(T) = \max\{|\lambda| : \lambda \in NR(T)\}$ $(= \sup\{|\lambda| : \lambda \in SNR(T)\})$.
4. We set $\mu(T) = \sup \mathcal{R}e\, SNR(T)$ $(= \max \mathcal{R}e\, NR(T))$.
5. $\rho(T)$ denotes the spectral radius of T; i.e., $\rho(T) = \lim_{n \to \infty} \|T^n\|^{1/n}$.

6a. T is said to be skew-Hermitian if $\mathcal{R}e\, NR(T) = \{0\}$ (i.e., if $\mathcal{R}e\, SNR(T) = \{0\}$).

b. T is said to be Hermitian if B is complex and $\mathcal{I}m\, NR(T) = \{0\}$, (i.e., if $NR(T)$ is real).

7. The Lie algebra of B, denoted by $\mathcal{A}(B)$ (or $\mathcal{A}$ when B is understood), is defined as the set of skew-Hermitian operators on B.
8. $\mathcal{I} = \mathcal{I}(B)$ denotes the set of linear isometries of B.

For basic results, see [2]. We have deliberately chosen the most elementary definitions. Of course $NR(T)$ is defined in [2] as the image of T on the state space of $\mathcal{L}(B)$; i.e., $NR(T) = \{< s , T > : (s , I) \in \Pi(\mathcal{L}(B)\}$, and our definition is shown equivalent to this. The spectral radius is defined so as to make sense on a real Banach space. Of course if B is complex and $\sigma(T)$ denotes the spectrum of T, i.e., $\{\lambda \in \mathbb{C} : T - \lambda I$ is not invertible$\}$ then we have the classical result that $\rho(T) = \max\{|\lambda| : \lambda \in \mathbb{C}\}$. It follows trivially that if B is complex, then T is a Hermitian operator on B if and only if iT is skew-Hermitian. However we shall see that our emphasis on $\mathcal{A}$, i.e., the skew-Hermitian operators, leads directly to the structure of the group of isometries of B, i.e., of $\mathcal{I}(B)$; moreover $\mathcal{A}$ can be defined and used for real as well as complex Banach spaces. Here are some further facts: $SNR(T)$ is connected but not necessarily convex (although this is true if B is a Hilbert space).

Thus if B is real, $NR(T) = \overline{SNR(T)}$ and moreover $nr(T) = \max\{|\mu(T)|, |\mu(-T)|\}$. If B is complex, it is easily seen that $nr(T) = \sup\{|\mu(\alpha T)| : |\alpha| = 1, \alpha \in \mathbb{C}\}$. We shall show below the standard result that for complex B, one has $\sigma(T) \subset SNR(T)$, hence via complexification we obtain that $\rho(T) \leq nr(T)$ for any real or complex Banach space. Finally we note the following deep result of Sinclair (cf. [2]): If T is in the linear span of the identity I and $\mathcal{A}(B)$, then $\|T\| = \rho(T)$ $(= nr(T))$. (For real spaces B, this follows from complexification; see §2, Prop. 2.6 below.) (The fact that $\|T\| = \rho(T)$ for Hermitian operators is due independently to Browder and Sinclair.)

We next recall some basic facts regarding the computation of $\mu(T)$. (We do not prove these here; the reader is referred to [2] or to [6], where a self-contained simple proof is given.)

<u>Theorem 1.1</u>. <u>Let</u> $T \in \mathcal{L}(B)$. <u>Then</u> $\mu(T) = \lim_{x \to 0+} \frac{\|I + xT\| - 1}{x}$

$$= \inf_{x > 0} \frac{\|I + xT\| - 1}{x} = \lim_{x \to 0+} \frac{\log\|e^{xT}\|}{x} = \sup_{x > 0} \frac{\log\|e^{xT}\|}{x}.$$

(Of course e^T is defined as $\sum_{n=0}^{\infty} \frac{T^n}{n!}$, where $T^0 = I$ and $0! = 1$ by convention).

The next result follows easily from the final equalities. (Both this <u>and</u> the preceding hold for either real or complex spaces.)

<u>Corollary 1.2</u>.

(a) $\mu(T)$ <u>is the smallest real number</u> μ <u>so that</u> $\|e^{xT}\| \leq e^{x\mu}$ <u>for all</u> $x > 0$.

(b) $nr(T)$ <u>is the smallest real number</u> r <u>so that</u> $\|e^{\lambda T}\| \leq e^{|\lambda| r}$ <u>for all scalars</u> λ.

We next present a simple proof that $nr(T)$ and $\|T\|$ are equivalent on complex Banach spaces. (For real spaces B, $T \in \mathcal{A}$ if and only if $nr(T) = 0$. Thus most of this article is devoted to the positive consequences of the <u>failure</u> of the equivalence of $nr(T)$ and $\|T\|$ on many real spaces.)

<u>Proposition 1.3</u>. (cf. [2].) <u>Let</u> B <u>be a complex Banach space and</u> $T \in \mathcal{L}(B)$. <u>Then</u> $\|T\| \leq e \cdot nr(T)$.

<u>Proof</u>: We make free use of elementary results concerning Banach-space valued analytic functions. Define $f : \mathbb{C} \to \mathcal{L}(B)$ by $f(z) = e^{zT}$ for all complex z. Then f is analytic. We first observe that $nr(T) = 0$ implies $T = 0$. Indeed, if $nr(T) = 0$, then by Corollary 1.2(b), $\|f(z)\| \leq 1$; hence by Liouville's theorem, $f(z)$ is constant so $f'(z)$ is identically zero. But $f'(z) = Te^{zT}$ so $0 = f'(0) = T$.

Next, we observe that our <u>definition</u> of $nr(T)$ shows that $T \to nr(T)$ is a semi-norm on $\mathcal{L}(B)$; we have just established that this is a <u>norm</u>. Hence by homogeneity it suffices to establish the desired inequality for T with $nr(T) = 1$. Now by the Cauchy integral formula,

$$T = f'(0) = \frac{1}{2\pi i} \int_\Gamma \frac{f'(\zeta)}{\zeta^2} d\zeta$$

where $\Gamma = \{z : |z| = 1\}$.

Hence

$$\|T\| \leq \frac{1}{2\pi} \int_\Gamma \|f'(\zeta)\| d|\zeta| . \tag{1}$$

But by Corollary 1.2(b),

$$\|e^{\lambda T}\| \leq e^{|\lambda|} \quad \text{for all complex } \lambda , \tag{2}$$

so $\|e^{\lambda T}\| \leq e$ for all $\lambda \in \Gamma$. Hence (1) and (2) yield that

$$\|T\| \leq \frac{1}{2\pi} \int_\Gamma e \, d|\zeta| = e ,$$

completing the proof.

We next present the fundamental connection between $A(B)$ and $I(B)$. (In fact for B finite-dimensional, $I(B)$ is a Lie-group and $A(B)$ is precisely its Lie algebra in the classical sense.)

<u>Theorem 1.4</u>. <u>Let</u> $T \in \mathcal{L}(B)$. <u>The following are equivalent</u>.

(a) $T \in A(B)$.

(b) $\|e^{xT}\| \leq 1$ <u>for all real</u> x .

(c) $e^{xT} \in I(B)$ <u>for all real</u> x .

(d) T <u>belongs to the tangent space of</u> $I(B)$ <u>at</u> I .

(e) $\lim_{x \to 0} \frac{\|I + xT\| - 1}{x} = 0$.

(For (d), we endow $I(B)$ with the <u>norm-topology</u> in $\mathcal{L}(B)$; of course $I(B)$ is a compact Lie group if B is finite-dimensional.) We thus interpret (d) to mean that there is a function $f : [-1 , 1] \to \mathcal{L}(B)$, <u>valued</u> in $I(B)$, differentiable at 0 , with $f(0) = I$ and $f'(0) = T$.

<u>Proof</u>: (a) ⇒ (b). Since $T \in A(B)$, $\mathcal{R}e\, NR(T) = \{0\}$ by definition, hence $\mu(T) = \mu(-T) = 0$. It follows by Corollary 1.2(a) that $\|e^{xT}\| \leq e^{x \cdot 0} = 1$ for all real x , establishing (b).

(b) ⇒ (c). Fix x real and set $U = e^{xT}$. Then $U^{-1} = e^{-xT}$. Since $\|U\|, \|U^{-1}\| \leq 1$, $U \in I(B)$.

(c) ⇒ (d). Define $f : [-1, 1] \to \mathcal{I}(B)$ by $f(x) = e^{xT}$. Thus f is differentiable and hence $T = f'(0)$ is in the tangent-space of $\mathcal{I}(B)$.

(d) ⇒ (e). Let f be as in the sentence directly preceding the proof of 1.4. Since $T = \lim_{x \to 0} \frac{f(x) - f(0)}{x}$ and $f(0) = I$,

$$f(x) = xT + I + o(x) . \tag{3}$$

(The symbol "$o(x)$" denotes some function $g(x)$ with $\lim_{x \to 0} \frac{g(x)}{x} = 0$.) Since f is valued in $\mathcal{I}(B)$, it follows from (3) that

$$1 = \|f(x)\| = \|xT + I\| + o(x) . \tag{4}$$

Of course (4) is equivalent to (e).

(e) ⇒ (a). We have by Theorem 1.1 that

$$\mu(T) = \lim_{x \to 0+} \frac{\|I + xT\| - 1}{x} = 0 , \tag{5}$$

$$\left\{ \begin{aligned} \mu(-T) &= \lim_{x \to 0+} \frac{\|I - xT\| - 1}{x} \\ &= -\lim_{x \to 0-} \frac{\|I + xT\| - 1}{x} = 0 . \end{aligned} \right. \tag{6}$$

(5) and (6) yield that $\mathcal{R}e\, NR(T) = \{0\}$ so $T \in \mathcal{A}(B)$. This completes the proof of Theorem 1.4.

We pass now to some permanence properties of the Lie algebra. Let us introduce a suggestive terminology; suppose X is a closed linear subspace of B, and there exists an operator $P : B \to X$ with $\|P\| = 1$ and $P|X = I_X$; (i.e., P is a norm-one linear projection onto X). For $T \in \mathcal{L}(B)$, we <u>define</u> $PT|X$ to be the <u>compression</u> of T to X (via P). Fix B a Banach space.

<u>Proposition 1.5</u>. <u>Let</u> $S, T \in \mathcal{A}$ <u>and</u> $U \in \mathcal{I}$. <u>Assume also</u> X <u>is a subspace of</u> B <u>and</u> $P : B \to X$ <u>is a norm-one surjective projection</u>.

(a) $U^{-1}TU$ <u>belongs to</u> $\mathcal{A}$.

(b) <u>The compression of</u> T <u>to</u> X <u>(via</u> P<u>)</u> <u>belongs to</u> $\mathcal{A}(X)$.

(c) $\mathcal{A}$ <u>is a weak-operator closed real linear space</u>.

(d) $TS - ST$ <u>belongs to</u> $\mathcal{A}$.

<u>Remark</u>: Of course (d) shows that $\mathcal{A}$ is <u>closed</u> under the canonical Lie-bracket operation. (d) is equivalent to the known result that if S and T are Hermitian operators on a complex space, then $i(ST - TS)$ is Hermitian.

<u>Proof</u>: It suffices to show these assertions for the case of real Banach

spaces, since if B is a complex Banach space, then an operator T belongs to $A(B)$ if and only if it belongs to $A(B_{\mathbb{R}})$ where $B_{\mathbb{R}}$ denotes B regarded as a real linear space. (Assuming B is real results only in notational simplification anyway.)

Now suppose $(b^*, b) \in \Pi$. It follows then easily that $((U^{-1})^* b^*, Ub) \in \Pi$. Hence $< b^*, U^{-1}TUb > = < (U^{-1})^* b^*, T(Ub) > = 0$, proving (a). ($V^*$ denotes the Banach space adjoint of an operator V.) Suppose $(x^*, x) \in \Pi(X)$. Now regarding $P : B \to X$, we have that $(P^*x^*, x) \in \Pi(B)$. Indeed, $\|P^*x^*\| \leq 1$, but $< P^*x^*, x > = < x^*, Px >$ $= < x^*, x > = 1$, showing this assertion. Hence $< x^*, PTx >$ $= < P^*x^*, Tx > = 0$, proving (b). (c) is an immediate consequence of our <u>definition</u> of A.

To prove (d), define $f : \mathbb{R} \to \mathcal{L}(B)$ by $f(t) = e^{-tS}Te^{tS}$ for all real t. Now for each real t, e^{tS} belongs to $\mathcal{I}$ by Theorem 1.4(c), hence $e^{-tS}Te^{tS} \in A$ by (a) of 1.5. Since A is norm-closed by (c), $f'(t) \in A$ for all real t. But $f'(t) = -Se^{-tS}Te^{tS} + e^{-tS}TSe^{tS}$, hence $f'(0) = -ST + TS$ belongs to A.

§2. Complex structures on real Banach spaces

Suppose X is a complex Banach space and let $X_{\mathbb{R}}$ denote X regarded as a Banach space over the real numbers. The main problem treated here: Given a real Banach space B, when is there a complex Banach space X so that B is isometric to $X_{\mathbb{R}}$?

<u>Definition</u>. <u>An operator</u> T <u>on</u> B <u>is said to induce a complex structure on</u> B <u>provided</u> $T^2 = -I$ <u>and</u> $aI + bT \in \mathcal{I}(B)$ <u>for all real numbers</u> a <u>and</u> b <u>with</u> $a^2 + b^2 = 1$.

It is easily seen that B is isometric to $X_{\mathbb{R}}$ for some complex X if and only if there is an operator T on B inducing a complex structure. Indeed, if for example $B = X_{\mathbb{R}}$, we simply define T by $Tb = ib$; then of course T induces a complex structure. Conversely suppose T induces a complex structure. Define multiplication over the complex numbers on B by $(\alpha + i\beta)b = \alpha b + T\beta b$ for all real α and β. The fact that $T^2 = -I$ shows easily that $w(zb) = (wz)b$ for all complex w and z. Since B is a real Banach space, to verify that $\|zb\| = |z|\|b\|$ for all complex z, it suffices to show this for $|z| = 1$, which of course follows immediately from the assumption that $aI + bT$ is norm-preserving for all real a and b with $a^2 + b^2 = 1$. Obviously the definition can be weakened in certain easy directions; for example it is evident that T induces a complex structure if and only if $T^2 = -I$ and $\|aI + bT\| \leq 1$ for all real a and b with $a^2 + b^2 = 1$. The main result of this section gives equivalent formulations in terms of the Lie algebra; its proof shows for example that T induces a complex structure provided $aI + bT$ is an isometry for all (a, b) sufficiently close to $(0, 1)$ with $a^2 + b^2 = 1$. The demonstration of this particular formulation seems to lie rather below the surface.

Theorem 2.1. Let B *be a real Banach space and* $T \in \mathcal{L}(B)$. *The following are equivalent:*

(a) T *induces a complex structure on* B.

(b) $T \in \mathcal{A}$ *and* $T^2 = -I$.

(c) $T \in \mathcal{A}$ *and* T *is an isometry*.

(d) T *is an isometry and* $\|aI + bT\| = 1$ *for all reals* a *and* b *with* $a^2 + b^2 = 1$.

Remarks.

1. It is not assumed *a priori* in (c) and (d) that $T \in \mathcal{I}$; that is, it is only assumed that T is norm-preserving.
2. This result was discovered independently of the considerably earlier work of Bollobás [1]; Theorem 2.1 is a light improvement on his work.
3. Our considerations here are isometric. The isomorphic theory is much simpler. Thus, suppose $T \in \mathcal{L}(B)$ and $T^2 = -I$. Define a new norm $|||\cdot|||$ on B by

$$|||x||| = \sup\{\|(aI + bT)(x)\| : a, b \text{ real with } a^2 + b^2 = 1\}.$$

Then $|||\cdot|||$ is equivalent to the original norm $\|\cdot\|$ and our discussion yields that T induces a complex structure on $(B, |||\cdot|||)$.

To prove Theorem 2.1, we note that (a) $\Rightarrow$ (d) is trivial. To see that (d) $\Rightarrow$ (c), define $f : \mathbb{R} \to \mathcal{L}(B)$ by $f(\theta) = \cos\theta I + \sin\theta T$. Then $f'(\theta) = -\sin\theta I + \cos\theta T$, so $f(0) = I$ and $f'(0) = T$. It follows that

$$f(x) = xT + I + o(x).$$

Hence

$$1 = \|f(x)\| = \|xT + I\| + o(x).$$

Of course this is equivalent to the assertion that $\lim_{x \to 0} \frac{\|xT + I\| - 1}{x} = 0$ so $T \in \mathcal{A}$ by Theorem 1.4(e).

The implication (c) ⇒ (b) lies below the surface, and we delay its proof, presenting first the demonstration of the remaining implication (b) ⇒ (a). We require the following simple result.

Lemma 2.2. Let $T \in \mathcal{L}(B)$. Then the following are equivalent:

(i) $T^2 = -I$.

(ii) $e^{xT} = \cos xI + \sin xT$ for all real x.

Proof of 2.2: Assuming first $T^2 = -I$, then $T^3 = -T$, $T^4 = I$, $T^5 = T$, etc., hence

$$\begin{aligned} e^{xT} &= I + xT - \frac{x^2}{2!} I - \frac{x^3}{3!} T + \frac{x^4}{4!} I + \frac{x^5}{5!} T - \frac{x^6}{6!} I - \frac{x^7}{7!} T + \ldots \\ &= I\left(1 - \frac{x^2}{2!} + \frac{x^4}{4!} - \frac{x^6}{6!} + \ldots\right) + T\left(x - \frac{x^3}{3!} + \frac{x^5}{5!} - \frac{x^7}{7!} + \ldots\right) \\ &= I(\cos x) + T(\sin x). \end{aligned}$$

Assuming (ii) holds, $e^{\frac{\Pi}{2}T} = T$ and $(e^{\frac{\Pi}{2}T})^2 = e^{\Pi T} = -I$, hence (i) holds.

Of course assuming (b), we now obtain (a) immediately from Lemma 2.2 and Theorem 1.4(c).

To handle the implication (c) ⇒ (b), we need the following result (cf. [2]).

Lemma 2.3. Let T be an operator on a complex Banach space X. Then the spectrum of T is contained in the closure of the spatial numerical range of T.

(Actually, we only require this for Hermitian operators, but the proof is simple enough in the general case, so we include it.)

Proof: Suppose $\lambda \notin \overline{SNR(T)}$; choose $\delta > 0$ so that

(7) $$|\lambda - z| > \delta \quad \text{for all} \quad z \in SNR(T).$$

Now let $b \in X$ with $\|b\| = 1$ and choose $b^* \in B^*$ with $(b^*, b) \in \Pi(x)$. Then

(8) $$\begin{aligned} \|(T - \lambda I)(b)\| &\geq |< b^*, (T - \lambda I)b >| \\ &= |< b^*, Tb > - \lambda| > \delta \quad \text{by (7).} \end{aligned}$$

(8) shows that $T - \lambda I$ is one-one with closed range; to see that $T - \lambda I$ is invertible and thus $\lambda \notin \sigma(T)$, we need only show that $T - \lambda I$ is an onto map. If not, there exists a $b^* \in B^*$ with $\|b^*\| = 1$ and b^* vanishing on the range of $T - \lambda I$, which implies that

(9) $$(T^* - \lambda I)b^* = 0.$$

By the Bishop-Phelps theorem, there exist $(b_n^*, b_n) \in \Pi$ with $b_n^* \to b^*$ in norm. Hence by (9), we have that

(10) $$\|(T^* - \lambda I)b_n^*\| \to 0 \quad \text{as} \quad n \to \infty .$$

But (10) yields that

(11) $$< (T^* - \lambda I)b_n^*, b_n > = < b_n^*, (T - \lambda I)b_n > \to 0 \quad \text{as} \quad n \to \infty .$$

Setting $z_n = < b_n^*, Tb_n >$, then $z_n \in SNR(T)$ for all n and (11) shows that $z_n \to \lambda$ as $n \to \infty$, a contradiction.

We prove (c) $\Rightarrow$ (b) by reducing to the analogous statement for Hermitian operators on complex Banach spaces. This result is essentiall due to Bollobás [1].

Proposition 2.4. Let S be a Hermitian isometry on a complex Banach space Z. Then $S^2 = -I$.

Proof: By the preceding result, $\sigma(S)$ is real since evidently $NR(S)$ is real. We next observe that S is surjective, i.e., $0 \notin S$. Otherwise, since the spectrum of S equals its own boundary (as a subset of $\mathbb{C}$), 0 would be an approximate eigenvalue, contradicting the assumption that S is an isometry. Thus it follows that

$$\sigma(S) \subset \mathbb{R} \cap \{z \in Z : |z| = 1\} = \{-1, 1\} .$$

Hence by the functional calculus, there exist closed linear subspaces X and Y of Z with $Z = X \oplus Y$, X, Y each invariant under S, and $\sigma(S|X) = \{1\}$, $\sigma(S|Y) = \{-1\}$. To complete the proof of 2.4, it suffices to show

Lemma 2.5. Let U be a Hermitian operator with $\sigma(U) = \{1\}$. Then $U = I$.

Indeed, once this is done, by the spectral mapping theorem we have that if U is Hermitian with $\sigma(U) = \{-1\}$, then $U = -I$; hence since $S|X$, $S|Y$ are Hermitian operators on X and Y respectively, we have $S|X = I_X$ and $S|Y = -I_Y$ whence $S^2 = I$.

To see Lemma 2.5, we let $V = U - I$; then evidently V is also Hermitian and $\sigma(V) = \{0\}$. By Sinclair's theorem $\|V\| = \rho(V) = 0$, hence $V = 0$ so $U = I$.

We now assume B is a real Banach space and introduce $\tilde{B}$, the standard complexification of B, as follows: Set $\tilde{B} = B \oplus B$. Define $\|\cdot\|$ on $\tilde{B}$ by

(12) $$\|x \oplus y\| = \sup_\theta \|\cos\theta x + \sin\theta y\| \quad \text{for all} \quad x, y \in B .$$

Define $i : \widetilde{B} \to \widetilde{B}$ by

(13) $$i(x \oplus y) = -y \oplus x \quad \text{for all} \quad x, y \in B.$$

It then follows that $\widetilde{B}$ is a complex Banach space under the norm given by (12) and complex multiplication induced by (13); of course B is then naturally isometrically embedded as the "real" part $B \oplus \{0\}$ of $\widetilde{B}$; we identify B with $B \oplus \{0\}$.

Definition. Given $T \in \mathcal{L}(B)$, define $\widetilde{T} : \widetilde{B} \to \widetilde{B}$ by $\widetilde{T}(x \oplus y) = Tx \oplus Ty$ for all $x, y \in B$.

The next fairly evident result summarizes all the needed permanence properties of this correspondence. (The algebraic motivation is that if T is a (real) linear operator on B, then $\widetilde{T}$ is a (complex) linear operator on $\widetilde{B}$.)

Proposition 2.6. Let $T \in \mathcal{L}(B)$.

(a) The map $T \to \widetilde{T}$ defined above is an algebraic isometry from $\mathcal{L}(B)$ into $\mathcal{L}(\widetilde{B})$ and $\widetilde{I_B} = I_{\widetilde{B}}$.

(b) Given $U \in \mathcal{L}(\widetilde{B})$, then $U = \widetilde{V}$ for some $V \in \mathcal{L}(B)$ if and only if B is invariant under U.

(c) T is invertible if and only if $\widetilde{T}$ is invertible.

(d) T is an isometry if and only if $\widetilde{T}$ is an isometry. Hence $T \in \mathcal{I}(B)$ if and only if $\widetilde{T} \in \mathcal{I}(\widetilde{B})$.

(e) $T \in \mathcal{A}(B)$ if and only if $\widetilde{T} \in \mathcal{A}(\widetilde{B})$.

The implication (c) $\Rightarrow$ (b) of Theorem 2.1 now follows from Propositions 2.4 and 2.6. Indeed, suppose T is an isometry with $T \in \mathcal{A}(B)$. Then $\widetilde{T}$ is an isometry with $\widetilde{T} \in \mathcal{A}(\widetilde{B})$. Hence $i\widetilde{T}$ is a Hermitian isometry, so $(i\widetilde{T})^2 = -(\widetilde{T})^2 = \widetilde{I}$, thus $\widetilde{T}^2 = -\widetilde{I} = (T^2)^{\sim}$, so $T^2 = -I$.

The proof of 2.6 is straightforward. (By (a) of 2.6, we mean that the mapping is linear and if $S, T \in \mathcal{L}(B)$, then $(ST)^{\sim} = \widetilde{S}\widetilde{T}$ and $\|\widetilde{T}\| = \|T\|$.) (a), (b) and (c) follow easily from the definitions. To see (d), suppose T is an isometry and let $x, y \in B$. Then

$$\begin{aligned} \|\widetilde{T}(x \oplus y)\| &= \sup_\theta \|\cos\theta Tx + \sin\theta Ty\| \\ &= \sup_\theta \|T(\cos\theta x + \sin\theta y)\| \\ &= \sup_\theta \|\cos\theta x + \sin\theta y\| \\ &= \|x \oplus y\| . \end{aligned}$$

Of course since B is invariant under $\widetilde{T}$, the converse assertion is trivial. The second assertion of (d) now follows immediately from the first and (c). Finally, to see (e), let x be a real number. Since our correspondence is algebraic, $(e^{xT})^{\sim} = e^{x\tilde{T}}$. Moreover by 2.6(d), $e^{xT} \in \mathcal{I}(B)$ iff $(e^{xT})^{\sim} \in \mathcal{I}(\tilde{B})$. Thus by Theorem 1.4, $T \in A(B)$ iff $e^{xT} \in \mathcal{I}(B)$ for all real x iff $e^{x\tilde{T}} \in \mathcal{I}(\tilde{B})$ iff $\tilde{T} \in A(\tilde{B})$, completin our proof. ("iff" means "if and only if".)

Remarks.

1. Of course it follows from 2.6 that for $T \in \mathcal{L}(B)$, $\sigma(T) = \sigma(\tilde{T}) \cap \mathbb{R}$. Moreover $\rho(T) = \lim_{n \to \infty} \|T^n\|^{1/n}$ $= \lim_{n \to \infty} \|\tilde{T}^n\|^{1/n} = \rho(\tilde{T}) = \sup\{|\lambda| : \lambda \in \sigma(\tilde{T})\}$. Thus Sinclair's theorem for complex spaces extends to real spaces also via 2.6.
2. Is there a proof of (c) ⇒ (b) of Theorem 2.1 which makes no use of complex numbers? All the rest of the arguments in this article do not require the complex scalars (except for Proposition 1.3).
3. It follows from the above proof that if T is an invertible member of $A(B)$ such that T and T^{-1} are power bounded, then again T induces a complex structure on B. Indeed, letting $S = i\tilde{T}$ as above, it follows that S is Hermitian and S, S^{-1} are power bounded; this again implies $\sigma(S) \subset \{-1, 1\}$ so our proof yields that $S^2 = I$ and hence $T^2 = -I$ as before.

§3. Some classification results

In this section we deal mostly with finite-dimensional real Banach spaces. We characterize Euclidean spaces in terms of the dimensions of their Lie algebras, classify all one-dimension B. Lie subalgebras up to orthogonal equivalence and obtain some results concerning rotation spaces.

Recall from the introduction that B is defined to be a rotation space if B is a functional unconditional sum of a Hilbert space H and $\mathbb{R}$ with $\dim H \geq 2$. Then B is isometric to $(H \oplus \mathbb{R})_{\underline{u}}$ for some 1-unconditional basis $\underline{u} = (u_1, u_2)$ of a two-dimensional Banach space U. If $U = (\mathbb{R}^2, \|\cdot\|)$ and (u_1, u_2) the unit vectors of $\mathbb{R}^2$, then the unit ball of B is obtained by rotating the ball of U about the x-axis, hence the terminology. Also recall: A Banach space is called Euclidean if it is isometric to a Hilbert space.

Our first result gives the isometric classification of 2- and 3-dimensional Banach spaces in terms of their Lie algebras.

<u>Theorem 3.1</u>. <u>Let</u> B <u>be a finite-dimensional real Banach space with Lie-algebra</u> $\mathcal{A}$.

(a) <u>If</u> $\dim B = 2$, <u>then</u> B <u>is Euclidean if and only if</u> $\mathcal{A} \neq \{0\}$.

(b) <u>If</u> $\dim B = 3$, <u>then</u> B <u>is a non-Euclidean rotation space if and only if</u> $\dim \mathcal{A} = 1$.

(c) <u>If</u> $\dim B = 3$, <u>then</u> B <u>is Euclidean if and only if</u> $\dim \mathcal{A} \geqslant 2$ (<u>if and only if</u> $\dim \mathcal{A} = 3$).

We shall present a combined algebraic-geometric proof of this result. The ideas developed will then prepare us for the following result which lies somewhat deeper.

<u>Theorem 3.2</u>. <u>Let</u> B <u>be as in</u> 3.1 <u>and</u> $n = \dim B$.

(a) B <u>is Euclidean if and only if</u> $\dim \mathcal{A} \geqslant \frac{(n-1)(n-2)}{2} + 1$ (<u>if and only if</u> $\dim \mathcal{A} = \frac{n(n-1)}{2}$).

(b) <u>If</u> B <u>is a non-Euclidean rotation space, then</u> $\dim \mathcal{A} = \frac{(n-1)(n-2)}{2}$.

<u>Remark</u>: In [7] we prove the converse to (b); that is, if $\dim \mathcal{A} = \frac{(n-1)(n-2)}{2}$, then B is a non-Euclidean rotation space.

The proofs of the assertions of 3.1 become transparent via the results of Section 1 and the following concept and its simple consequences.

<u>Definition</u>. Let B be a Banach space isomorphic to a Hilbert space, with norm $\|\cdot\|$. A norm $\|\cdot\|_H$ is called a <u>compatible</u> norm on B if $\|\cdot\|$ and $\|\cdot\|_H$ are equivalent and setting $H = (B, \|\cdot\|_H)$, then

(a) H is Euclidean

and

(b) $\mathcal{I}(B) \subset \mathcal{I}(H)$.

The next result follows easily from Theorem 1.4; it is of course well known.

<u>Proposition 3.3</u>. <u>Let</u> $\|\cdot\|_H$ <u>be a compatible norm on</u> B <u>and let</u> $\langle , \rangle$ <u>be an inner product inducing</u> $\|\cdot\|_H$.

(a) $\mathcal{A}(B) \subset \mathcal{A}(H)$.

(b) $T \in \mathcal{A}(H)$ <u>if and only if</u> $T^* = -T$; <u>i.e.</u>, T <u>is skew-Hermitian in the classical sense</u>.

<u>Proof</u>: To see (a), let $T \in \mathcal{A}(B)$ and x a real number. Then $e^{xT} \in \mathcal{I}(B)$ by 1.4(a), hence $e^{xT} \in \mathcal{I}(H)$ so $T \in \mathcal{A}(H)$.

For (b), we have that $\Pi(H)$ may simply be identified with the sphere of H itself, where for $y \in H$, we set $f_y(x) = \langle x, y \rangle$. Thus, $T \in A(H)$ if and only if $\mathcal{R}e \langle Tx, x \rangle = 0$ for all x with $\|x\| = 1$; hence for arbitrary x, expanding $\langle T(x+y), x+y \rangle$ and $\langle Tix + y, ix + y \rangle$, one obtains easily that this occurs if and only if $\langle Tx, y \rangle = -\langle T^*x, y \rangle$ for all $x, y \in H$.

Now suppose B is real n-dimensional. Then as is well-known, B has a compatible norm. (We shall sketch a proof momentarily, for completeness.) We may then reinterpret things as follows: B may be identified with $(\mathbb{R}^n, \|\cdot\|)$ for some norm $\|\cdot\|$; letting $\|\cdot\|_E$ be the usual norm on $\mathbb{R}^n$, we have that $\|\cdot\|_E$ is compatible with $\|\cdot\|$. Hence $\mathcal{I}(B) \subset \mathbb{O}_n = \mathcal{I}(\mathbb{R}^n)$. (Here $\mathbb{O}_n$ denotes the orthogonal group of real $n \times n$ matrices; also $S\mathbb{O}_n$ denotes the "special orthogonal group" of orthogonal matrices of determinant one. We identify $n \times n$ matrices on $\mathbb{R}^n$ with $\mathcal{L}(\mathbb{R}^n)$.) Hence $A(B) \subset A(\mathbb{R}^n) = so_n$, the set of $n \times n$ - skew-symmetric matrices. (Of course exactly the same considerations apply for $\mathbb{C}^n$.) Thus $A(B)$ is a Lie-subalgebra of so_n, under the natural bracket $[S, T] = ST - TS$, as proved in Proposition 1.5(d). Now suppose $\|\cdot\|$ and $\|\cdot\|_E$ are compatible and suppose $U \in \mathbb{O}_n$. Define a new norm $\|\cdot\|'$ on $\mathbb{R}^n$ by $\|x\|' = \|Ux\|$ for all $x \in \mathbb{R}^n$. Then it is evident that $\|\cdot\|_E$ is also compatible with $\|\cdot\|'$ and $T \in A(B')$ iff $U^*TU \in A(B)$ (where $B' = (\mathbb{R}^n, \|\cdot\|')$; $U^* = U^{-1}$ since U is orthogonal).

<u>Definition</u>. Subsets $\mathfrak{S}$ and $\mathfrak{T}$ of $\mathcal{L}(\mathbb{R}^n)$ are orthogonally equivalent if there is a $U \in \mathbb{O}_n$ with $U^* \mathfrak{S} U = \mathfrak{T}$.

Our above comments show that if $\mathfrak{T}$ is orthogonally equivalent to $A'(B)$, then $\mathfrak{T}$ equals $A(B')$ for some B' isometric to B. The converse is also true.

<u>Proposition 3.4</u>.

(a) *Let B be an n-dimensional Banach space. Then B has a compatible norm.*

(b) *Let B and B' be isometric n-dimensional spaces. Assume $B = (\mathbb{R}^n, \|\cdot\|)$ and $B' = (\mathbb{R}^n, \|\cdot\|')$ so that $\|\cdot\|_E$ is compatible with both $\|\cdot\|$ and $\|\cdot\|'$. Then $A(B)$ is orthogonally equivalent to $A(B')$.*

<u>Remark</u>: The content of (b) is thus: The family of equivalence classes of B. Lie-subalgebras of so_n, under orthogonal equivalence, provides a set of isometric invariants for n-dimensional Banach spaces.

<u>Proof of</u> (a): It is well-known that this may be achieved by the Fritz-John ellipsoid. However I prefer the more algebraic proof (also well-

known). $\mathcal{I}(B)$ is a compact group under multiplication, hence has a unique Haas-measure m. That is, m is a probability measure on the Borel subsets of $\mathcal{I}(B)$ so that $m(TS) = m(S)$ for all Borel sets S and $T \in \mathcal{I}(B)$. Let $(,)$ be an inner product on B and define $<,>$ on B by

$$< x , y > = \int_{\mathcal{I}(B)} (Ux , Uy)\,dm(U) \quad \text{for all } x , y \in B .$$

It follows easily that $<,>$ is also an inner product on B. If $V \in \mathcal{I}(B)$ and $x , y \in B$, then $< Vx , Vy > = < x , y >$ by invariance of m under left multiplication, hence $V \in \mathcal{I}(H)$ where $H = (B , \|\cdot\|_H)$ and $\|\cdot\|_H$ is the norm induced by $<,>$.

Proof of (b): This is a simple consequence of a result in group-representation theory. Let $G = \mathcal{I}(B)$ and $G' = \mathcal{I}(B')$. Hence, G and G' are both closed subgroups of $\mathcal{O}_n$. Let $T : B' \to B$ be a linear isometry. It follows that of course T is a one-one onto map and $T^{-1}GT = G'$. That is, G and G' are similar. By a classical result, G and G' are unitarily equivalent; in our setting, this means orthogonally equivalent. (For completeness, we sketch a proof below.) Hence there is a $U \in \mathcal{O}_n$ with $U^*CU = G'$. It follows easily, using for example the criterion of Theorem 1.4(c), that $U^* A(B)U = A(B')$, thus $A(B)$ and $A(B')$ are orthogonally equivalent.

Here is a proof of the classical result stated above. Suppose first U and V belong to $\mathcal{O}_n$ and R is a positive definite matrix with $R^{-1}UR = V$. Then we claim $U = V$ (and moreover R commutes with U). Indeed, $(R^{-1}UR)(RU^*R^{-1}) = VV^* = I$. Hence $UR^2U^* = R^2$ or $UR^2 = R^2U$. Since R is a function of R^2, being positive definite, this implies $UR = RU$, so $U = V$. Now let G and G' be similar subgroups of $\mathcal{O}_n$, and choose T an invertible matrix with $T^{-1}GT = H$. By polar decomposition we may choose U in $\mathcal{O}_n$ and R positive definite with $T = UR$. Now setting $\widetilde{G} = U^*GU$, G and $\widetilde{G}$ are orthogonally equivalent. But then $G' = R^{-1}\widetilde{G}R = \widetilde{G}$ by our considerations above; that is, G' and G are orthogonally equivalent.

Remark: The following is a rather famous open problem: Suppose B is a separable Banach space with $\mathcal{I}(B)$ transitive on the sphere $S(B)$ of B. (That is, given x , y in B of norm 1, there is a $T \in \mathcal{I}(B)$ with $Tx = y$.) Is B isometric to a Hilbert space? Now suppose B is isomorphic to a Hilbert space and suppose in fact there exists a compatible norm $\|\cdot\|_H$ on B. Then the answer is affirmative. Indeed, fix x_0 in $S(B)$ and assume without loss of generality $\|x_0\|_H = 1$. Then if x in $S(B)$, choose $T \in \mathcal{I}(B)$ with $Tx_0 = x$.

Since $T \in \mathcal{I}(H)$, $\|x\|_H = 1$. Thus $\|\cdot\|_H = \|\cdot\|$, so B is isometric to $H = (B, \|\cdot\|_H)$. (This proof holds for non-separable B as well.) Hence in particular, we obtain the known result that the answer is affirmative for finite-dimensional B. (As far as I know, this problem is open for B an isomorph of Hilbert space, separable or not. There are known counter examples in the non-separable setting.)

<u>From now on, an n-dimensional space</u> B <u>shall be assumed equal to</u> $(\mathbb{R}^n, \|\cdot\|)$ <u>with</u> $\|\cdot\|_E$ <u>and</u> $\|\cdot\|$ <u>compatible</u>.

We next proceed with Theorem 3.1. It is useful to set up notation for the sequel. Let R_0 denote $90°$ counter-clockwise rotation in $\mathbb{R}^2$. Thus the operator R_0 is (represented by) the matrix $\begin{bmatrix} 0 & -1 \\ 1 & 0 \end{bmatrix}$. Let $R(\theta)$ denote counter-clockwise rotation by θ radians. Thus $R(\theta)$ is (represented by) the matrix $\begin{bmatrix} \cos\theta & -\sin\theta \\ \sin\theta & \cos\theta \end{bmatrix}$. By Lemma 2.2,

$$e^{\theta R_0} = \cos\theta I + \sin\theta R_0 = R(\theta) . \tag{14}$$

In particular, (14) shows of course that $e^{\frac{\Pi}{2} R_0} = R_0$.

Let us now consider Theorem 3.1. In case (a), if B is $(\mathbb{R}^2, \|\cdot\|)$ then $T \in \mathcal{A}(B)$ if and only if (the matrix for) T is skew-symmetric; i.e., of the form $\begin{bmatrix} 0 & -c \\ c & 0 \end{bmatrix}$ for some scalar c; whence $\mathcal{A}(B)$ is one-dimensional. Assume conversely that $\mathcal{A}(B)$ is 1-dimensional; let (e_1, e_2) be the usual basis for $\mathbb{R}^2$; i.e., $e_1 = (1, 0)$ and $e_2 = (0, 1)$. We may assume that $\|e_1\| = \|e_1\|_E = 1$ by multiplying $\|\cdot\|$ by a constant if necessary. Since $\mathcal{A}(B) \subset \mathcal{A}(\mathbb{R}^2)$, we may choose $T \in \mathcal{A}(B)$ with $T = R_0$. Hence $e^{\theta T} = R(\theta) \in \mathcal{I}(B)$ and hence $\|R(\theta)e_1\| = 1$ for all θ. We have thus proved that the <u>sphere</u> of B contains the sphere of $\mathbb{R}^2$; i.e., $x \in \mathbb{R}^2$, $\|x\|_E = 1 \Rightarrow \|x\| = 1$. Of course this proves that the balls of B and $\mathbb{R}^2$ coincide, whence $B = (\mathbb{R}^2, \|\cdot\|_E)$. (In a sense, the result is an immediate consequence of 3.4(a), by the above remark.)

Before proceeding, let us draw an immediate consequence of this argument.

<u>Corollary 3.5</u>. <u>Let</u> B <u>be an arbitrary real Banach space and</u> T <u>a rank-2 norm-one member of</u> $\mathcal{A}(B)$. <u>Then</u> $X = TB$ <u>is Euclidean and</u> $T|X$ <u>is an isometry</u>. <u>Moreover if</u> e <u>is a norm-one member of</u> X, <u>then</u> (e, Te) <u>is equivalent to the Euclidean basis</u>. <u>That is</u>, $\|\alpha e + \beta Te\| = (\alpha^2 + \beta^2)^{\frac{1}{2}}$ <u>for all scalars</u> α <u>and</u> β.

Next suppose B is a non-Euclidean 3-dimensional rotation space. We may choose a basis (e_1, e_2, e_3) for B so that $X = [e_1, e_2]$ is isometric to $\mathbb{R}^2$ and moreover if U is an isometry of X, then $x + ye_3 \to Ux + ye_3$ is an isometry of B. That is, for all θ,

$\begin{bmatrix} R(\theta) & 0 \\ 0 & 1 \end{bmatrix}$ is the matrix for an isometry of B. ($\begin{bmatrix} R(\theta) & 0 \\ 0 & 1 \end{bmatrix}$ denotes the matrix $\begin{bmatrix} \cos\theta & -\sin\theta & 0 \\ \sin\theta & \cos\theta & 0 \\ 0 & 0 & 1 \end{bmatrix}$.) Since $\begin{bmatrix} R(\theta) & 0 \\ 0 & 1 \end{bmatrix} = e^{\theta R_0}$, R_0 is the matrix of an element of $A(B)$, so $\dim A(B) \geq 1$.

We prove later that $\dim A(B) = 1$. Suppose next that $\dim A(B)=1$. Choose $T \in A(B)$, $T \neq 0$. Since T is a skew-symmetric matrix, rank $T = 2$; (or else T would have a real non-zero eigenvalue, contradicting its skew-symmetry). We may thus choose an orthonormal basis for $\mathbb{R}^3$ so that the matrix for T with respect to this basis is of the form $\begin{bmatrix} 0 & -c & 0 \\ c & 0 & 0 \\ 0 & 0 & 0 \end{bmatrix}$ for some $c \neq 0$. Assume $c = 1$. We may in fact also <u>assume</u> the basis is the standard Euclidean basis for $\mathbb{R}^3$ and again suppose $\|e_1\| = 1$. By Corollary 3.5, we obtain that if $X = [e_1, e_2]$, then $\|\cdot\| = \|\cdot\|_E$ on X. Then it follows that $e^{\theta T} = \begin{bmatrix} R(\theta) & 0 \\ 0 & 1 \end{bmatrix}$ for all θ. But this implies that B is a functional unconditional sum of $\mathbb{R}^2$ and $\mathbb{R}$. Indeed suppose $x, x' \in X$ with $\|x\| = \|x'\|$. Choose θ so that $R(\theta)x = x'$. Then for any real y, $\|x + y\| = \|e^{\theta T}(x + y)\| = \|x' + y\|$; hence also $\|x + y\| = \|-x + y\|$ $= \|x - y\| = \|x' - y\|$. Thus B is a rotation space.

Finally, suppose $\dim A(B) \geq 2$. We shall show that then B is Euclidean, completing the proof of all assertions (it's trivial that $\dim A(\mathbb{R}^3) = 3$). We proceed exactly as above; thus assume $T = \begin{bmatrix} R_0 & 0 \\ 0 & 0 \end{bmatrix} \in A(B)$. Now suppose $S \in A(B)$ with S and T linearly independent. We may then choose V a linear combination of S and T so that

$$V = \begin{bmatrix} 0 & 0 & a \\ 0 & 0 & b \\ -a & -b & 0 \end{bmatrix}$$

with $a^2 + b^2 = 1$. Suppose $u = ae_1 + be_2$. Then $\|u\| = 1$, $u \in VB$ and $\|Vu\| = 1$. Since $V \in A(B)$, it follows by Corollary 3.5 that $(u, Vu) = (u, -e_3)$ is isometrically equivalent to the Euclidean basis for $\mathbb{R}^2$. Finally, suppose scalars (x_1, x_2, x_3) given and set $\alpha = (x_1^2 + x_2^2)^{\frac{1}{2}}$. Then since B is a rotation space, $\|x_1e_1 + x_2e_2 + x_3e_3\| = \|\alpha u + x_3e_3\| = (\alpha^2 + x_3^2)^{\frac{1}{2}} = (x_1^2 + x_2^2 + x_3^2)^{\frac{1}{2}}$, proving that $B = (\mathbb{R}^3, \|\cdot\|_E)$.

<u>Remark</u>. A more algebraic proof of Theorem 3.1(c) goes as follows: Having obtained the vectors T and V as above, a computation shows that $TV - VT$ is <u>independent</u> of T and V. Hence $\dim A(B) \geq 3$ but $A(B) \subset so_3$ whence $A(B) = so_3$ so $e^{A(B)} = e^{so_3} = SO_3$ by a standard result. Thus the sphere of B contains the sphere of $\mathbb{R}^3$ as before, proving that $B = (\mathbb{R}^3, \|\cdot\|_E)$. We have chosen the somewhat more geometric proof to develop intuition for later results.

We delay the proof of 3.2 for awhile, and proceed now to determine those finite-dimensional Banach spaces with a non-trivial Lie algebra, and also to classify the one-dimensional Lie algebras of Banach spaces, up to orthogonal equivalence. All our results here are simple consequences of known (but sometimes quite deep!) classical theorems.

We first need a standard result from linear algebra.

Lemma 3.6. *Let A be a non-zero $n \times n$ skew-symmetric real matrix. Then A has even rank and A is orthogonally equivalent to a direct sum of multiples of rotation matrices. That is, there exist k, positive reals $c_1, \ldots, c_k$ and U an orthogonal matrix so that $U^*AU = B$, where*

$$B = \begin{bmatrix} c_1R_0 & & & & \\ & c_2R_0 & & & \\ & & \ddots & & \\ & & & c_kR_0 & \\ & & & & 0 \end{bmatrix} .$$

Perhaps the quickest proof is to unitarily diagonalize A over the complex numbers and then obtain the desired orthogonal transformation by working with the real and complex parts of the complex eigenvectors of A. This argument works also for real normal matrices and shows incidentally that if A and B are real normal matrices, then A unitarily equivalent to B implies A orthogonally equivalent to B.

Henceforth, we fix $B = (\mathbb{R}^n, \|\cdot\|)$ an n-dimensional real space. Terms such as "orthogonal," "orthonormal basis" refer to $(\mathbb{R}^n, \|\cdot\|_E)$ under the standard inner product $\langle x, y \rangle = \sum_{i=1}^n x_iy_i$ if $x = (x_1 \ldots, x_n)$ and $y = (y_1, \ldots, y_n)$. Now Lemma 3.6 and our preceding results give the complete description of the structure of a single element of $\mathcal{A}(B)$.

Corollary 3.7. *Let T be a non-zero operator on B.*

(a) *Suppose there exists an orthonormal basis for B, and real numbers $c_1, \ldots, c_k$ such that if A is the matrix for T with respect to this basis, then*

$$A = \begin{bmatrix} c_1R_0 & & \\ & \ddots & \\ & & c_kR_0 \end{bmatrix} .$$

Then T is in $\mathcal{A}(B)$ if and only if

$$\begin{bmatrix} R(c_1\theta) & & & \\ & \ddots & & \\ & & R(c_k\theta) & \\ & & & I \end{bmatrix}$$

is the matrix of an operator in $\mathcal{I}(B)$ with respect to this basis, for all θ.

(b) Suppose then T is in $A(B)$. Then there is a k so that rank $T = 2k$ and positive reals $c_1, \ldots, c_k$, and an orthonormal basis $(u_1, v_1, u_2, v_2, \ldots, v_k, v_k, w_1, \ldots, w_\ell)$ so that the matrix for T for this basis is as above (where $\ell = n - 2k$). It follows that setting $E_i = [u_i, v_i]$, then $E_1, \ldots, E_k$ are orthogonal, the null space of T equals the orthogonal complement of $E_1 + \ldots + E_k$, and for each i, E_i is invariant under T and E_i is isometric to $\mathbb{R}^2$ (in fact $\|\cdot\|$ is a multiple of $\|\cdot\|_E$ on E_i).

Proof of (a): Since $e^{xR_0} = R(x)$, we have that

$$e^{\theta A} = \begin{bmatrix} R(c_1\theta) & & & \\ & \ddots & & \\ & & R(c_k\theta) & \\ & & & I \end{bmatrix}$$

for all θ; of course the matrix for $e^{\theta T}$ equals $e^{\theta A}$; hence this equivalence follows immediately from Theorem 1.4.

Part (b) is an immediate consequence of Lemma 3.6 and Theorem 3.1(a). (The c_j's may be chosen positive since R_0 is orthogonally equivalent to $-R_0$.)

The next result is a trivial consequence of the above and deep classical results.

Theorem 3.8. The following are equivalent:

(a) B has a non-trivial Lie algebra.

(b) B has infinitely many isometries.

(c) There exists a 2-dimensional Euclidean subspace E of B so that every isometry of E extends to an isometry of B.

Proof: (a) ⇒ (c). This follows immediately from Corollary 3.7.
(c) ⇒ (b). Since $\mathcal{I}(E)$ is infinite, so is $\mathcal{I}(B)$.
(b) ⇒ (a). It follows by compatibility of our norm that $\mathcal{I}(B) \subset \mathcal{I}(\mathbb{R}^n) = \mathcal{O}_n$. In fact, $\mathcal{I}(B)$ is thus an infinite closed subgroup of $\mathcal{O}_n$,

hence is a Lie group (cf. [3]). Thus $\mathcal{I}(B)$ has a non-zero tangent space at I, whence by Theorem 1.4, $\mathcal{A}(B) \neq 0$.

We wish next to characterize one-dimensional Lie algebras of finite-dimensional Banach spaces. We have need of an elegant result of Robbin, determining those Lie subalgebras of so_n which are a B. Lie algebra.

Definitions. Let $\mathcal{A}$ be a subset of $\mathcal{L}(\mathbb{R}^n)$.

(a) $\mathcal{A}$ is a Lie-subalgebra of so_n if $\mathcal{A}$ is a linear subspace of so_n such that $ST - TS \in \mathcal{A}$ for all $S, T \in \mathcal{A}$.

(b) $\mathcal{A}$ is a B. Lie algebra if there is a norm $\|\cdot\|$ on $\mathbb{R}^n$ with $\|\cdot\|$ and $\|\cdot\|_E$ compatible so that $\mathcal{A} = \mathcal{A}(B)$ where $B = (\mathbb{R}^n, \|\cdot\|)$. (Thus by Proposition 3.3, $\mathcal{A}$ is then a Lie-subalgebra of so_n.)

(c) $\mathcal{A}$ is saturated if for all $T \in \mathcal{L}(\mathbb{R}^n)$, if $Tx \in \mathcal{A}x$ for all $x \in \mathbb{R}^n$, then $T \in \mathcal{A}$; (where $\mathcal{A}x = \{Ax : A \in \mathcal{A}\}$).

We set $e^{\mathcal{A}} = \{e^A : A \in \mathcal{A}\}$.

Theorem 3.9. Let $\mathcal{A}$ be a Lie subalgebra of so_n. Then the following are equivalent.

(a) $\mathcal{A}$ is a B. Lie algebra.

(b) $\mathcal{A}$ is saturated and $e^{\mathcal{A}}$ is a compact group (under multiplication and the standard topology on $\mathcal{L}(\mathbb{R}^n)$).

The first part of the implication of (a) ⇒ (b) is trivial. The second part is quite deep; in fact we obtain (by standard results in the theory of Lie groups) that $e^{\mathcal{A}}$ is the component of the identity of $\mathcal{I}(B)$. For (b) ⇒ (a), one obtains rather easily that $\mathcal{A}$ is then the Lie-algebra of the compact group $e^{\mathcal{A}}$ (which I prefer to think of as simply the tangent space to $e^{\mathcal{A}}$ at I).

The remaining implication follows from Robbin's result [5], after disposing of a somewhat subtle point. (We drop temporarily our convention that $\|\cdot\|_E$ and $\|\cdot\|$ are compatible.) Suppose we say that $\|\cdot\|_H$ is weakly compatible with $\|\cdot\|$ if $\mathcal{I}_0(B) \subset \mathcal{I}(H)$ where $\mathcal{I}_0(B)$ denotes the component of the identity of $\mathcal{I}(B)$ and $H = (B, \|\cdot\|_H)$ is Euclidean. Since $\mathcal{A}(B)$ is the tangent space of $\mathcal{I}_0(B)$ at I, we have that if $\|\cdot\|_H$ is weakly compatible with $\|\cdot\|$, then still $\mathcal{A}(B) \subset \mathcal{A}(H)$. Now Robbin's result shows that if (b) is satisfied, then setting $G = e^{\mathcal{A}}$, there is a norm $\|\cdot\|$ on $\mathbb{R}^n$ so that setting $B = (\mathbb{R}^n, \|\cdot\|)$, then $G = \mathcal{I}_0(B)$. Since $e^{\mathcal{A}} \subset e^{so_n} = \mathcal{I}_0(\mathbb{R}^n)$, we have that $\|\cdot\|_E$ is weakly compatible with $\|\cdot\|$. But now suppose $\|\cdot\|$

is another norm on $\mathbb{R}^n$, with $\|\cdot\|_E$ compatible with $\|\cdot\|'$ and $B' = (\mathbb{R}^n, \|\cdot\|')$ isometric to B. Then letting $G' = I_0(B')$, we have as in the proof of 3.4(b) that G' and G are similar, hence again orthogonally equivalent. Thus $A = A(B)$ and $A(B')$ are orthogonally equivalent as before. Thus there is indeed a norm $|||\cdot|||$ on $\mathbb{R}^n$ with $\|\cdot\|_E$ and $|||\cdot|||$ compatible and $A = A(\widetilde{B})$ where $\widetilde{B} = (\mathbb{R}^n, |||\cdot|||)$. Hence A is a B. Lie algebra.

We now resume our convention that $B = (\mathbb{R}^n, \|\cdot\|)$ with $\|\cdot\|_E$ and $\|\cdot\|$ compatible. The next result yields a simple classification of one-dimensional Banach Lie algebras up to orthogonal equivalence.

<u>Definition</u>. Fix k and n positive integers with $2k \leq n$ and set $\Lambda_k^n = \{(u_1, \dots, u_k) : 1 \leq u_1 \leq u_2 \dots \leq \dots \leq u_k$ are positive integers with $\{u_1, \dots, u_k\}$ relatively prime$\}$. For each $u \in \Lambda_k^n$, let $T_u = u_1R_0 \oplus \dots \oplus u_kR_0$; that is, T_u is the skew-symmetric matrix

$$\begin{bmatrix} u_1R_0 & & & & \\ & u_2R_0 & & & \\ & & \ddots & & \\ & & & u_kR_0 & \\ & & & & 0 \end{bmatrix}.$$

Then let $A_u = [T_u]$, the one-dimensional span of the operator T_u.

<u>Theorem 3.10</u>. <u>Let</u> B <u>be an</u> n-<u>dimensional space with a one-dimensional Lie algebra</u>. <u>Then there is a unique</u> k <u>and</u> $u \in \Lambda_k^n$ <u>with</u> $A(B)$ <u>orthogonally equivalent to</u> A_u. <u>Conversely given any</u> $u \in \Lambda_k^n$, <u>there exists a</u> B n-<u>dimensional with</u> $A_u = A(B)$.

<u>Remark</u>: The interest from isometric Banach space theory is thus: If $u \neq u'$, $A(B) = A_u$ and $A(B') = A_{u'}$, then B and B' are not isometric. The first non-trivial case occurs in $\mathbb{R}^4$. We then obtain that $\Lambda_1^4 = \{(1)\}$ as always; $\Lambda_2^4 = \{(1, 1)\} \cup \{(k, n) : 1 \leq k \leq n$ and k, n are relatively prime integers$\}$. Thus there are infinitely many non-equivalent one-dimensional B. Lie subalgebras of so_4.

<u>Proof</u>: Suppose first B has a one-dimensional Lie algebra. Then let $T \in A(B)$, $T \neq 0$. By Corollary 3.7, there exist k reals with $0 < c_1 \leq c_2 \dots \leq c_k$ so that T is orthogonally equivalent to $c_1R_0 \oplus \dots \oplus c_kR_0$ (and thus $\operatorname{rank} T = 2k \leq n$). Since e^A is compact, we must have (c_i, c_j) pairwise linearly dependent over the rational numbers. Otherwise by a classical result, $e^A = \{R(\theta c_1) \oplus \dots \oplus R(\theta c_k) \oplus I : \theta$ real$\}$ would not be closed. We may thus assume without loss of generality, since A is linear, that $(c_1, \dots, c_k)$ are integers. But <u>now</u> if j equals the greatest common divisor of $\{c_1, \dots, c_k\}$

and $u_i = \frac{c_i}{j}$ for all j, then $\{u_1, \dots, u_k\}$ is relatively prime and of course $\mathbb{A}$ is spanned by T_u where $u = (u_1, \dots, u_k)$.

Now suppose $v \in \Lambda_i^n$ for some i and $\mathbb{A}_v$ orthogonally equivalent to $\mathbb{A}_u$ as above. Since a non-zero member of $\mathbb{A}_v$ has rank $2i$ and one of $\mathbb{A}_u$ has rank $2k$, $i = k$. We then obtain that for some non-zero t, T_u is similar to T_{tv}. In fact this implies $u = tv$, which in turn yields that $t = 1$ (with $t = \frac{a}{b}$, a, b relatively positive integers; b must divide v_i all i, hence $b = 1$, and a must divide v_i all i, hence $b = 1$).

Finally, suppose $u \in \Lambda_k^n$ and set $\mathbb{A} = \mathbb{A}_u$. If m = least common multiple of $\{u_1, \dots, u_k\}$, then $e^{\mathbb{A}} = \{R(\theta u_1) \oplus \dots \oplus R(\theta u_k) \oplus I : 0 \leq \theta \leq \frac{2\Pi}{m}\}$ is a compact group. To verify Robbin's condition, we need only check that $\mathbb{A}$ is saturated. Let A an $n \times n$ matrix and assume $Ax \in \mathbb{A}x$ for all $x \in \mathbb{R}^n$. Since so_n is saturated, $A \in so_n$. Letting $x \in \mathbb{R}^n$ and for $1 \leq i \leq k$, x_i the projection of x in $Y_i = [e_{2i-1}, e_{2i})$ (where $(e_1, \dots, e_n)$ is the natural basis of $\mathbb{R}^n$) we thus have: Given x, there is a real t_x so that $Ax = t_x(u_1R_0x_1 \oplus \dots \oplus u_kR_0x_k)$. It certainly follows that Y_i is invariant under A and hence $A|Y_i$ is skew-Hermitian. Moreover $Ax = 0$ if x is in the orthogonal complement of $Y_1 + \dots + Y_k$. Thus choose reals $d_1, \dots, d_k$ so that $A = d_1R_0 \oplus \dots \oplus d_kR_0$. Now let x be a fixed vector with $x_i \neq 0$ for all i, $1 \leq i \leq k$. We thus obtain that $t_x(u_1R_0x_1 \oplus \dots \oplus u_kR_0x_k) = d_1R_0x_1 \oplus \dots \oplus d_kR_0x_k$ which implies $d_j = t_xu_j$ for all j; thus $A = t_xT_u$, so $A \in \mathbb{A}$.

<u>Remark</u>. Actually Theorem 5.3 is an explicit working out of a special case of a result of Robbin [5], based on earlier work of Schneider and Turner [8], although these authors work exclusively with complex numbers. In fact, their work really characterizes all commutative B. Lie algebras. Equivalently it characterizes those subgroups of so_n which are commutative, connected, and arise as the component of the identity of the group of isometries of an n-dimensional Banach space. Thus one obtains the following result: Let $\mathbb{A}$ be a commutative subspace of so_n. Then $\mathbb{A}$ is a B. Lie algebra if and only if $\mathbb{A}$ is orthogonally equivalent to an algebra $\mathbb{A}'$ where $\mathbb{A}'$ has a basis, each term of which has the form $c_1R_0 \oplus \dots \oplus c_kR_0$, with the c_i's <u>integers</u> and $2k = n$. (That is, if A is a basis-element, then

$$A = \begin{bmatrix} c_1R_0 & & \\ & \ddots & \\ & & c_kR_0 \\ & & & 0 \end{bmatrix},$$

the c_i's as above.)

We pass now to the proof of Theorem 3.2. As always, we assume $B = (\mathbb{R}^n, \|\cdot\|)$ with $\|\cdot\|_E$ and $\|\cdot\|$ compatible. We first need a result about rank-2 members of $A(B)$, which will prove useful in the sequel. This result is really implicit in our proof of Theorem 3.1(a).

Proposition 3.11. *Let T be a rank-2 member of $A(B)$, let X equal the range of T and Y the orthogonal complement of X (with respect to $\|\cdot\|_E$). Then there is a $c > 0$ so that $\|x\| = c\|x\|_E$ for all $x \in X$. Y is the kernel of T. If $U \in \mathcal{I}(X)$, then $U \oplus I \in \mathcal{I}(B)$. If x and x' are in X with $\|x\| = \|x'\|$ and $y \in Y$, then $\|x + y\| = \|x' + y\| = \|x - y\|$.*

Proof: The first assertion follows immediately from the proof of 3.1(a). The second one follows because $T \in so_n$. By multiplying T by a non-zero constant, we may assume the matrix for $T|X$ equals R_0 with respect to some orthonormal basis for X. Then if $U \in \mathcal{I}_0(X)$, there is a real c so that $U = e^{cT|X}$; then evidently $U \oplus I = e^{cT}$. The last assertion is an immediate consequence of the preceding and the transitivity of $\mathcal{I}_0(X)$ on the sphere of X. In turn, this implies $U \oplus I \in \mathcal{I}(B)$ if $U \in \mathcal{I}(X)$.

We also need an elementary result about Euclidean spaces; we leave the proof to the reader. (The result is essentially the observation that one-unconditional bases of $\mathbb{R}^2$ are isometrically equivalent to the usual basis.)

Proposition 3.12. *Let W be a Euclidean Banach space and $x, y \in W$ be such that $\|\alpha x + \beta y\| = \|\alpha x - \beta y\|$ for all scalars α and β. Then $\|x + y\| = (\|x\|^2 + \|y\|^2)^{\frac{1}{2}}$.*

Let us now fix some notation. Let $\mathcal{B} = (e_1, \ldots, e_n)$ be an orthonormal basis for $\mathbb{R}^n$. For $T \in \mathcal{L}(\mathbb{R}^n)$, we let $[T]_{\mathcal{B}}$ denote the matrix for T with respect to $\mathcal{B}$ (when $\mathcal{B}$ is understood, we set $[T] = [T]_{\mathcal{B}}$). Let $Q_{\mathcal{B}} = Q$ denote the orthonormal projection onto $Z_{\mathcal{B}} = Z = [e_2, \ldots, e_n]$.

Define a map $\alpha_{\mathcal{B}} = \alpha$ by

$$(15) \qquad \alpha : A(B) \to so_{n-1}, \quad \alpha(T) = [QTQ] \quad \text{for all } T \in A(B) .$$

(The range of α is indeed contained in so_{n-1} since $A(B) \subset so_n$.) Of course α is a linear map.

We prove 3.2(a) by induction on $\dim B$. The result is proved if $\dim B = 2$; so we fix $n > 2$ and assume the result proved for all spaces with dimension less than or equal to $n-1$. We now assume $\dim A(B) > \frac{(n-1)(n-2)}{2}$.

Lemma 3.13. There is a T in $A(B)$ with $\operatorname{rank} T = 2$.

Proof: Fix B an orthonormal basis and let $\alpha = \alpha_B$ as in (17). Any non-zero T in the null space of α is rank-2. Were this false, α would be one-one and hence onto. This is impossible because then $\dim \alpha(A(B)) = \dim so_{n-1} = \frac{(n-1)(n-2)}{2}$.

We now need a basic principle about Lie subalgebras of so_n. We shall delay its proof until later.

Theorem 3.14. Let A be a Lie subalgebra of so_n and assume that A contains all matrices A of the form

$$A = \begin{bmatrix} 0 & a_2 \dots a_n \\ -a_2 & \\ \vdots & \quad 0 \\ -a_n & \end{bmatrix}.$$

Then $A = so_n$.

Now choose T a rank-2 member of $A(B)$, let $X = TB$ and $Y = X^{\perp}$ (the orthogonal complement of X). Let $B = (e_1, \dots, e_n)$ be an orthonormal basis for $\mathbb{R}^n$ with $X = [e_1, e_2]$ and $Y = [e_3, \dots, e_n]$. Let $Q = Q_B$. We let $n(\alpha)$ = nullity of α = dim null space of α, $r(\alpha)$ = rank α = dim range α.

Corollary 3.15. If $n(\alpha) = n-1$, then B is Euclidean.

Proof: We have that if $A = \{[T] : T \in A(B)\}$, then A satisfies the hypotheses of 3.14, hence $A = so_n$.

By a standard result in linear algebra, $e^{so_n} = SO_n$, so in fact $\|\cdot\|$ is a multiple of $\|\cdot\|_E$ since SO_n is transitive on the sphere of $\mathbb{R}^n$.

Lemma 3.16. $\|Q\| = 1$.

Proof: Let $b = \alpha e_1 + \beta e_2 + y$ for scalars α and β and $y \in Y$. Then by Proposition 3.11,

$$(16) \qquad \|\alpha e_1 + \beta e_2\| = \|(\alpha^2 + \beta^2)^{\frac{1}{2}} e_2\|.$$

Setting $x = \alpha e_1 + e_2$ and $x' = (\alpha^2 + \beta^2)^{\frac{1}{2}} e_2$, we have by 3.11 that $\|x + y\| = \|x' + y\| = \|x' - y\|$, hence by (16),

(17) $$\|b\| = \|x' + y\| \geqslant \|\beta e_2 + y\| = \|Qb\| .$$

(The last inequality follows since in fact we now have that $([e_2], Y)$ is a one-unconditional decomposition of $Z = Z_{\mathcal{B}}$ (as defined before (15)).)

We now obtain that the range of our map $\alpha = \alpha_{\mathcal{B}}$ is <u>contained</u> in $\mathbb{A}(Z)$ (where we identify $\mathbb{A}(Z)$ here with $\{[T]_{\underline{\mathcal{B}}} : T \in \mathbb{A}(Z)$, $\underline{\mathcal{B}} = (e_2, \ldots, e_n)\}$).

Indeed, this follows immediately from the "compression" result Proposition 1.5(b).

<u>Lemma 3.17</u>. $Z_{\mathcal{B}}$ <u>is Euclidean</u>.

<u>Proof</u>: By inductive hypothesis, we need only establish that

(18) $$r(\alpha) > \frac{(n-2)(n-3)}{2} .$$

For then we have that $\dim \mathbb{A}(Z) > \frac{(n-2)(n-3)}{2}$, hence Z is Euclidean. If $n(\alpha) = n - 1$, we are done anyway, by Corollary 3.15. So assume $n(\alpha) < n - 1$. Then by the "rank plus nullity" theorem,

$$r(\alpha) + n(\alpha) = \dim \mathbb{A}(B) \geqslant \frac{(n-1)(n-2)}{2} + 1 ,$$

hence

$$r(\alpha) > \frac{(n-1)(n-2)}{2} + 1 - (n-1) = \frac{(n-2)(n-3)}{2} ,$$

establishing (18) and hence 3.17.

The completion of the proof of Theorem 3.2(a) (modulo Theorem 3.14) is as follows:

We have now established that X and Y are both Euclidean. To complete the proof, we need only show that

(19) $$\|x + y\| = (\|x\|^2 + \|y\|^2)^{\frac{1}{2}} \quad \text{for all } x \in X \text{ and } y \in Y .$$

So fix $x \in X$ and $y \in Y$. Of course we assume $x \neq 0$. Then we may assume by homogeneity that

$$\|x\|_E = 1 .$$

Now choose $e_1' \in x$, with $\|e_1'\|_E = 1$ and e_1' orthogonal to x; setting $e_2' = x$, we then have that $\mathcal{B}' = \{e_1', e_2', e_3, \ldots, e_n\}$ is an orthonormal basis for B. Thus by Lemma 3.17, $Z' = Z_{\mathcal{B}'}$ is Euclidean. Hence by the last equality of 3.11,

(20) $$\|\alpha x + \beta y\| = \|\alpha x - \beta y\| \quad \text{for all scalars } \alpha \text{ and } \beta .$$

Since x, y lie in the Euclidean space Z', (19) holds by Proposition 3.12, completing the proof of 3.1(a).

Remark. It is perhaps worth observing that if B is Euclidean and $\|\cdot\|_E$ and $\|\cdot\|$ are compatible, then in fact $\|\cdot\|$ is a multiple of $\|\cdot\|_E$. Indeed, fix x_0 in B and assume without loss of generality that $\|x_0\|_E = \|x_0\| = 1$. Since B is Euclidean, if $x \in B$, $\|x\| = 1$, there is a $T \in I(B)$ with $Tx_0 = x$. But then $T \in I(\mathbb{R}^n)$ as well; thus $\|x\|_E = 1$, so $\|\cdot\| = \|\cdot\|_E$.

It remains to prove Theorem 3.14. Let $[A, B] = AB - BA$. Let $(e_1, \dots, e_n)$ be the standard basis; for $i \neq j$, let $R(i, j)$ be the matrix so that $R(e_1) = e_j$, $R(e_j) = -e_i$, and $R(e_\ell) = 0$ if $\ell \neq i$ or j. Now the hypotheses of 3.14 imply that $R(e_1, e_j) \in A$ for all $j \neq 1$. A computation shows that if $i \neq 1$, $j \neq 1$, then

(21) $$[R(e_1, e_i), R(e_1, e_j)] = R(e_i, e_j).$$

Thus since A is a Lie algebra, we obtain that $R(e_i, e_j) \in A$ for all $i \neq j$. Thus 3.14 is established once we prove the following:

(22) $$\dim[R(e_i, e_j) : 1 \leq i, j \leq n, \ i \neq j] = \frac{n(n-1)}{2}.$$

We establish this by induction on n. The result is trivial for $n = 2$. Suppose $n > 2$ and the result proved for $n - 1$. Let $X = [R(e_1, e_j) : 1 < j \leq n]$, $Y = [R(e_i, e_j) : 1 < i, j \leq n, \ i \neq j]$, and $Z = [R(e_i, e_j) : 1 \leq i, j \leq n, \ i \neq j]$. Then $X + Y = Z$ and $X \cap Y = 0$. It's clear that $\dim X = n - 1$; $\dim Y = \frac{(n-1)(n-2)}{2}$ by inductive hypothesis; hence $\dim Z = \frac{(n-1)(n-2)}{2} + n - 1 = \frac{n(n-1)}{2}$, establishing (20).

Theorem 3.2(b) is a simple consequence of 3.2(a). Indeed, suppose B is a non-Euclidean rotation space. Now choose X a codimension-one Euclidean subspace and Y a one-dimensional subspace. (X, Y) is a functional unconditional decomposition of B. It follows that if $U \in I(X)$, then $U \oplus I \in I(B)$. Hence if $A \in A(X)$, $A \oplus 0 \in A(B)$, since $e^{(A \oplus 0)} = e^A \oplus I$. Hence $\dim A(B) \geq \dim A(X) = \frac{(n-1)(n-2)}{2}$. Equality follows by 3.2(a).

References.

1. B. Bollobás, "A property of Hermitian elements," J. London Math. Soc. (2) 4 (1971), 379-380.

2. F.F. Bonsall and J. Duncan, Numerical ranges of operators on normed spaces and of elements of normed algebras, vols. I (1971) and II (1974), Cambridge University Press.

3. S. Helgason, Differential Geometry and Symmetric Spaces, Academic Press, New York, 1962.

4. N.J. Kalton and G.V. Wood, "Orthonormal systems in Banach spaces and their applications," Math. Proc. Cambridge Philos. Soc. 79 (1976) No. 3, 493-510.

5. J.W. Robbin, "Lie algebras of infinitesimal norm isometries," Lin. Alg. and its Appl. 10 (1975), 95-102.

6. H. Rosenthal, "On one-complemented subspaces of complex Banach spaces with a one-unconditional basis, according to Kalton and Wood," Israel Seminar on Geometrical Aspects of Functional Analysis 1983-1984, University of Tel Aviv.

7. ____________, "Functional Hilbertian Sums," to appear.

8. H. Schneider snd R.E.L. Turner, "Norm Hermitian matrices," Lin. Alg. and its Appl. 8 (1974), 375-412.

WEAK*-DENTING POINTS IN DUALS OF OPERATOR SPACES

W.M. Ruess and C.P. Stegall
Department of Mathematics
University of Essen
Univ.str.3, 43 Essen 1
Fed. Rep. Germany

SUMMARY: We characterize the weak*-denting points in the dual unit balls of the spaces of compact, of weakly compact, and of all bounded linear operators between Banach spaces X and Y in terms of the denting points of the unit ball of X and the weak*-denting points of the dual unit ball of Y.

1. INTRODUCTION

The extremal structure of the dual unit ball B_{Z*} of a Banach space Z provides information on smoothness properties of the underlying space Z. Classical results by V.L. Smul'yan [11,12] (cf. [1]) particularly demonstrate that the wealth of extreme points, or of w*-exposed and w*-strongly exposed points of B_{Z*}, respectively, has a direct bearing on Gateaux- and Fréchet-differentiability properties of the norm on Z. Among the more recent observations in this direction is the result [4, Thm.2.1] that a Banach space Z has Mazur's intersection property (see the definition preceding Proposition 7 in section 2) if and only if the set of w*-denting points of the dual unit ball B_{Z*} is (norm) dense in the unit sphere S_{Z*}.

In our papers [9,10], we dealt with the problem of characterizing the extremal structure of dual unit balls of operator spaces in terms of the corresponding extremal structures of the (bi-) dual unit balls of the domain and range spaces, covering the cases of extreme, [w*-] exposed, and [w*-] strongly exposed points, respectively.

In this paper, we take up the particular case of w*-denting points in duals of operator spaces. If we agree to denote the spaces of all bounded, of all weakly compact, and of all compact linear operators from a Banach space X into a Banach space Y by L(X,Y), W(X,Y), and K(X,Y), respectively, then one of our main results (Theorem 4 in section 2) reads as follows:

$$\text{w*-dent } B_{K(X,Y)*} = \text{w*-dent } B_{W(X,Y)*} = \text{w*-dent } B_{L(X,Y)*} =$$
$$= \text{dent } B_X \otimes \text{w*-dent } B_{Y*},$$

where a tensor $x \otimes y^*$ acts on an operator $u \in L(X,Y)$ in the natural way: $x \otimes y^*(u) = (ux, y^*)$.

This result has two particular aspects: First, the weak*-denting points in the dual unit ball of any of these operator spaces are completely determined by the denting points of the unit ball of the domain space X and the weak*-denting points of the dual unit ball of the range space Y. Furthermore, although these operator spaces (and their duals) may differ quite a bit, their dual unit balls all have exactly the same set of weak*-denting points.

The object of this paper is to derive this and related results, and to consider some consequences. Parts of the results of this paper have been announced in section 5 of [8].

Notation and terminology: Throughout the paper, all linear spaces are assumed to be over the reals. Our Banach space terminology is standard. For a Banach space Z, B_Z denotes the closed unit ball of Z, and S_Z the unit sphere. For $z_0 \in Z$, and $\varepsilon > 0$, we denote by $B(z_0;\varepsilon)$ the closed ε-ball in Z with center z_0.

The sets of extreme points and of denting points of B_Z will be denoted by $\text{ext}\, B_Z$ and $\text{dent}\, B_Z$, respectively, and the set of weak*-denting points of B_{Z^*} by $w^*\text{-dent}\, B_{Z^*}$.

Recall that, for a norming linear subspace F of Z^*, a point $z_0 \in S_Z$ is said to be an F-denting point of B_Z, if, for every $\varepsilon > 0$, $z_0 \notin \sigma(Z,F)$-$\text{cl co}(B_Z \setminus B(z_0;\varepsilon))$ = closure of the convex hull of $B_Z \setminus B(z_0;\varepsilon)$ with respect to the weak topology on Z generated by $F \subset Z^*$. Particular cases are:

(a) $\text{dent}\, B_Z = Z^*\text{-dent}\, B_Z$, and (b) $w^*\text{-dent}\, B_{Z^*} = Z\text{-dent}\, B_{Z^*}$.

For equivalent definitions and the relevance of denting and weak*-denting points in the context of Radon-Nikodym properties, the reader is referred to [1,2,3 and 6].

Besides the familiar operator spaces mentioned above, we shall, for given Banach spaces X and Y, consider also the more general operator spaces $L_{w^*}(X^*,Y)$ [resp. $K_{w^*}(X^*,Y)$] of all weak*-weakly continuous [resp. compact and weak*-weakly continuous] linear operators from X^* into Y, both endowed with the usual operator norm. This will enable us to apply our results to specific operator spaces that are representable as (subspaces of) a suitable $L_{w*}(X^*,Y)$- or $K_{w*}(X^*,Y)$-space

(1.1) $$W(X,Y) = L_{w*}(X^{**},Y)$$
$$u \longrightarrow u^{**}$$

(1.2) $$K(X,Y) = K_{w*}(X^{**},Y)$$
$$u \longrightarrow u^{**}$$

(1.3) The completed injective tensor product $X \tilde{\otimes}_\varepsilon Y$ of X and Y is a closed linear subspace of $K_{w*}(X^*,Y)$, the (isometrical) linear embedding being given by

$$X \tilde{\otimes}_\varepsilon Y \longrightarrow K_{w*}(X^*,Y)$$
$$x \otimes y \longrightarrow [x^* \to x^*(x)y]$$

Note that $X \tilde{\otimes}_\varepsilon Y = K_{w*}(X^*,Y)$ whenever either of X and Y has the approximation property.

Finally, the space of bounded bilinear forms of the product $X \times Y$ of Banach spaces X and Y will be denoted by $B(X,Y)$.

2. RESULTS

Our results on weak*-denting points in duals of operator spaces will be based on the following two technical propositions.

1. Proposition: Assume that U and V are closed norming linear subspaces of X^* and Y^*, respectively, and let H be a linear subspace of the space $B(U,V)$ of bounded bilinear forms on $U \times V$, containing $X \otimes Y$: $X \otimes Y \subset H \subset B(U,V)$.

Then we have: if $x_0^* \in U \cap$ w*-dent B_{X^*} and $y_0^* \in V \cap$ w*-dent B_{Y^*}, then $x_0^* \otimes y_0^* \in$ w*-dent B_{H^*}.

Proof: According to the assumptions on x_0^* and y_0^*, we have:

There exist sequences $0 < \alpha_i < \beta_i < 1$, $\alpha_i \to 1$, and $0 < \varepsilon_i \to 0$, and H

(1) $(x_i)_i \subset S_X$ such that $x_0^*(x_i) > \beta_i$, and $B_{X^*} \cap (X_i > \alpha_i) \subset B(x_0;\varepsilon_i)$, and,

accordingly, sequesces $0 < \alpha_i' < \beta_i' < 1$, $\alpha_i' \to 1$, and $0 < \varepsilon_i' \to 0$, and $(y_i)_i \subset S_Y$ such that $y_0^*(y_i) > \beta_i'$, and $B_{Y^*} \cap (y_i > \alpha_i') \subset B(y_0; \varepsilon_i')$.

We conclude that, for all $n \in \mathbb{N}$

(2) $$x_0^* \otimes y_0^* \notin w^*\text{-clco}\Big(\bigcup_{i,j\le n} B_{H^*} \cap (x_i \otimes y_j \le \beta_i\beta_j')\Big).$$

Thus, there exist sequences $(h_n)_n \subset S_H$ and $0 < \rho_n < 1$ such that $h_n(x_0^*,y_0^*) > \rho_n > \sup_{h^*\in S_n} h^*(h_n)$, where $S_n = \bigcup_{i,j\le n} B_{H^*} \cap (x_i \otimes y_j \le \beta_i\beta_j')$.

Now, let $K_n = B_{H^*} \cap (h_n > \rho_n)$, and choose $h_n^* \in K_n$ for $n \in \mathbb{N}$. We have to show that $\|h_n^* - x_0^* \otimes y_0^*\|_{H^*} \to 0$.

Case 1: Assume that $H \subset K_{w^*}(X^*,Y)$.
According to [5,Ch. I, §4.1,Prop.18], there exist $(\lambda_n)_n \subset M^+(B_{X^*} \times B_{Y^*})$ with $\|\lambda_n\| = \|h_n^*\|$ such that

(3) $$h_n^*h = \int_{B_{X^*} \times B_{Y^*}} (hx^*,y^*)\,d\lambda_n(x^*,y^*).$$

As a first step, we prove that, for all $i \in \mathbb{N}$,

$$\lambda_n(B_{X^*} \cap (x_i > \alpha_i) \times B_{Y^*}) \longrightarrow 1 \quad \text{as } n \to \infty.$$

For, assume there exist $i_1 \in \mathbb{N}$ and $0 < \beta < 1$, and a sequence $(n_m)_m \subset \mathbb{N}$

(4) $$\lambda_{n_m}(B_{X^*} \cap (x_{i_1} > \alpha_{i_1}) \times B_{Y^*}) \le \beta \quad \text{for all } m \in \mathbb{N}.$$

We construct strictly increasing sequences $(i_k)_k$, $(j_k)_k$, $(m_k)_k$, $(\alpha_{i_k})_k$, $(\beta_{i_k})_k$, $(\beta_{j_k}')_k$ and $(n_{m_k})_k$ such that $\min\{i_{k+1},j_{k+1}\} > n_{m_k} > \max\{n_{m_{k-1}},i_k,j_k\}$, and

(5) $$\alpha_{i_k} < \beta_{i_k}\beta_{j_k}' \quad \text{for all } k \in \mathbb{N}.$$

First, there exists $j_1 \in \mathbb{N}$ such that $\alpha_{i_1} < \beta_{i_1}\beta_j'$ for all $j > j_1$. Choose $m_1 \in \mathbb{N}$ such that $n_{m_1} > \max\{i_1,j_1\}$. Then, choose $i_2 > \max\{i_1,n_{m_1}\}$ such that $\alpha_{i_1} < \alpha_{i_2}$ and $\beta_{i_1} < \beta_{i_2}$. Then, there exists $j_2 > \max\{j_1,n_{m_1}\}$ such that $\alpha_{i_2} < \beta_{i_2}\beta_j'$ for all $j > j_2$, and $\beta_{j_2}' > \beta_{j_1}'$. At this point choose $m_2 > m_1$ such that $n_{m_2} > \max\{n_{m_1}, i_2, j_2\}$. The induction procedure ought to be clear from here on.

We know that $\int x_i \otimes y_j d\lambda_{n_{m_k}} = h^*_{n_{m_k}}(x_i \otimes y_j) > \beta_i \beta'_j$ for all $i,j \le n_{m_k}$, for, otherwise, $h^*_{n_{m_k}} \in S_{n_{m_k}}$, contradicting the fact that $h^*_{n_{m_k}} \in K_{n_{m_k}}$. We conclude that

$$(6) \qquad \int x_{i_k} \otimes y_{j_k} d\lambda_{n_{m_k}} > \beta_{i_k} \beta'_{j_k} > \alpha_{i_k} \quad \text{for all } k \in \mathbb{N}.$$

This means that $(x_{i_k}, \int y^*(y_{j_k}) x^* d\lambda_{n_{m_k}}) > \alpha_{i_k}$, so that, for $x^*_k = \int y^*(y_{j_k}) x^* d\lambda_{n_{m_k}}$, we conclude from (1) that

$$(7) \qquad \|x^*_k - x^*_o\| \longrightarrow 0.$$

Now, let $E = (B_{X^*} \cap (x_{i_1} > \alpha_{i_1})) \times B_{Y^*}$, and E^c its complement in $B_{X^*} \times B_{Y^*}$, and let

$$a^*_k = (\lambda_{n_{m_k}}(E))^{-1} \int_E y^*(y_{j_k}) x^* d\lambda_{n_{m_k}}, \quad \text{and}$$

$$b^*_k = (\lambda_{n_{m_k}}(E^c))^{-1} \int_{E^c} y^*(y_{j_k}) x^* d\lambda_{n_{m_k}}, \quad \text{so that}$$

$$x^*_k = \lambda_{n_{m_k}}(E) a^*_k + \lambda_{n_{m_k}}(E^c) b^*_k .$$

Then $(\lambda_{n_{m_k}}(E))_k$ clusters at $r \ge 0$, $(\lambda_{n_{m_k}}(E^c))_k$ clusters at $s \ge 0$, $(a^*_k)_k$ clusters weak* at $a^*_o \in B_{X^*}$, and $(b^*_k)_k$ clusters weak* at $b^*_o \in B_{Y^*}$ so that, by (7): $x^*_o = r a^*_o + s b^*_o$, where, from (4) and (6), $0 \le r \le \beta < 1$ and $r + s = 1$. Since $x^*_o \in \text{ext} B_{X^*}$, we conclude that $x^*_o = b^*_o$. But, for some subnet $(b^*_{k_\gamma})_\gamma$ of $(b^*_k)_k$, we have:

$$x^*_o(x_{i_1}) = b^*_o(x_{i_1}) = \lim_\gamma b^*_{k_\gamma}(x_{i_1}) \le \alpha_{i_1}.$$

This contradiction shows that $\lambda_n(B_{X^*} \cap x_i > \alpha_i) \times B_{Y^*}) \to 1$ for all $i \in \mathbb{N}$. Analogous arguments show that $\lambda_n(B_{X^*} \times B_{Y^*} \cap (y_j > \alpha_j')) \to 1$ for all $j \in \mathbb{N}$.

We now show that $\|h_n^* - x_o^* \otimes y_o^*\|_{H^*} \to 0$.

Let $\varepsilon > 0$; according to what we have shown so far, there exist i_o, j_o and $n_o \in \mathbb{N}$ such that $\varepsilon_{i_o} < \varepsilon/8$, $\varepsilon'_{j_o} < \varepsilon/8$ and $\lambda_n(B_{X^*} \cap (x_{i_o} > \alpha_{i_o}) \times B_{Y^*}) > 1 - \varepsilon/8$, and $\lambda_n(B_{X^*} \times B_{Y^*} \cap (y_{j_o} > \alpha'_{j_o})) > 1 - \varepsilon/8$ for all $n \geq n_o$.

Let $F = (B_{X^*} \cap (x_{i_o} > \alpha_{i_o})) \times (B_{Y^*} \cap (y_{j_o} > \alpha'_{j_o}))$. Then $\lambda_n(F) > 1 - \varepsilon/4$, and, for any $h \in B_H$:

$$|h_n^* h - h(x_o^*, y_o^*)| = |\int h d\lambda_n - h(x_o^*, y_o^*)|$$

$$\leq |\int_F [(h(x^* - x_o^*), y^*) + (hx_o^*, y^* - y_o^*)] d\lambda_n| + |1 - \lambda_n(F)|\, |h(x_o^*, y_o^*)| +$$

$$+ |\int_{F^c} (hx^*, y^*) d\lambda_n| \leq \varepsilon_{i_o} + \varepsilon'_{j_o} + \varepsilon/4 + \varepsilon/4 < \varepsilon \quad \text{for all } n \geq n_o .$$

This completes the proof in case $H \subset K_{w^*}(X^*, Y)$.

<u>Case 2</u>: Assume that H is any linear subspace of $B(U,V)$, containing $X \otimes Y$: $X \otimes Y \subset H \subset B(U,V)$.

We shall show that $x_o^* \otimes y_o^* \in H\text{-dent } B_{U \tilde{\otimes}_\pi V}$. For then we have

$x_o^* \otimes y_o^* \in H\text{-dent } B_{(U \tilde{\otimes}_\pi V)^{**}} = H\text{-dent } B_{B(U,V)^*}$, and thus, a fortiori,

$x_o^* \otimes y_o^* \in w^*\text{-dent } B_{H^*}$.

Choose $(t_n)_n \subset B_{U \tilde{\otimes}_\pi V}$ with $(t_n, h_n) > \rho_n$. We may even assume that $\|t_n\|_\pi < 1$ and $(t_n, h_n) > \rho_n$.

Then $t_n = \sum_{i=1}^{\infty} \lambda_{n,i} x^*_{n,i} \otimes y^*_{n,i}$, where $\|x^*_{n,i}\| \leq 1$, $\|y^*_{n,i}\| \leq 1$, and $\lambda_{n,i} \geq 0$, $\sum_{i=1}^{\infty} \lambda_{n,i} \leq 1$.

Now, consider the measures $\lambda_n = \sum_{i=1}^{\infty} \lambda_{n,i} \delta_{(x^*_{n,i}, y^*_{n,i})}$.

According to what we have shown in the first case above, we know that

$$\lambda_n(\{(x^*,y^*) \in B_{X^*} \times B_{Y^*} \mid x^*(x_i) > \alpha_i \text{ and } y^*(y_j) > \alpha_j'\}) \to 1$$

for all $i,j \in \mathbb{N}$. Thus, given $\varepsilon > 0$, there exist i,j and $n_o \in \mathbb{N}$ such that $\varepsilon_i < \varepsilon/4$, $\varepsilon_j' < \varepsilon/4$, and

$$\lambda_n(\{(x^*,y^*) \in B_{X^*} \times B_{Y^*} \mid x^*(x_i) > \alpha_i \text{ and } y^*(y_j) > \alpha_j'\}) > 1 - \varepsilon/4$$

for all $n \geq n_o$.

Let $E = \{(x^*,y^*) \in B_{X^*} \times B_{Y^*} \mid x^*(x_i) > \alpha_i \text{ and } y^*(y_j) > \alpha_j'\}$, and, for $n \geq n_o$ $\sigma_n = \{m \in \mathbb{N} \mid (x^*_{n,m}, y^*_{n,m}) \in E\}$.

Then we have: $1 - \varepsilon/4 < \lambda_n(E) = \sum\limits_{\sigma_n} \lambda_{n,i}$.

Let $r_n = (\sum\limits_{\sigma_n} \lambda_{n,i}) x_o^* \otimes y_o^*$; then, for $n \geq n_o$:

$$\|r_n - x_o^* \otimes y_o^*\|_\pi = \|(1 - \sum_{\sigma_n} \lambda_{n,i}) x_o^* \otimes y_o^*\|_\pi = |1 - \sum_{\sigma_n} \lambda_{n,i}| < \varepsilon/4 ,$$

and

$$\|r_n - t_n\|_\pi \leq \sum_{i \in \sigma_n} \lambda_{n,i} \|x_o^* \otimes y_o^* - x^*_{n,i} \otimes y^*_{n,i}\|_\pi + \sum_{i \notin \sigma_n} \lambda_{n,i} <$$

$$< \varepsilon_i + \varepsilon_j' + \varepsilon/4 < (3/4)\varepsilon \quad \text{for all } n \geq n_o.$$

We thus conclude that $\|t_n - x_o^* \otimes y_o^*\|_\pi < \varepsilon$ for all $n \geq n_o$, which completes the proof of Proposition 1.

2. <u>Proposition</u>: Assume that U and V are closed norming linear subspaces of X^* and Y^*, respectively, and that H is a linear subspace of the space $B(U,V)$ of bounded bilinear forms on $U \times V$, containing $X \otimes Y$: $X \otimes Y \subset H \subset B(U,V)$. Let $U_H^* = \{h(\cdot, y^*) \mid y^* \in V\}$ and $V_H^* = \{h(x^*, \cdot) \mid x^* \in U\}$.

Then we have: If $T_o \in w^*\text{-dent } B_{H^*}$, then $T_o = x_o^* \otimes y_o^*$ for some $(x_o^*, y_o^*) \in U_H^*\text{-dent } B_U \times V_H^*\text{-dent } B_V$.

<u>Proof</u>: Assume that $T_o \in S_{H^*}$ is a weak*-denting point of B_{H^*}. Then we have:

For every $\varepsilon > 0$, there exist $0 < \alpha(\varepsilon) \leq \epsilon$ and $h(\varepsilon) \in S_H$

(1) such that (i) $T_o(h(\varepsilon)) > 1 - \alpha(\varepsilon)$, and

(ii) $B_{H^*} \cap (h(\varepsilon) > 1 - \alpha(\varepsilon)) \subset B(T_o;\varepsilon)$.

We first show that T_o is of the form $x_o^* \otimes y_o^*$ for some $(x_o^*,y_o^*) \in S_U \times S_V$.

Given $n \in \mathbb{N}$, let $h_n = h(1/n) \in S_H$ according to (1), and choose $(x_n^*)_n \subset S_U$ and $(y_n^*)_n \subset S_V$ such that $h_n(x_n^*,y_n^*) > 1 - \alpha(1/n)$. Then we have, by (1) (ii): $\|T_o - x_n^* \otimes y_n^*\|_{H^*} \leq 1/n$, so that $(x_n^* \otimes y_n^*)_n$ is uniformly Cauchy over $B_X \times B_Y$. At this point, we use Lemma 1.2 in section 1 of [10] to conclude that some subsequences $(x_{n_i}^*)_i$ and $(y_{n_i}^*)_i$ converge to some $x_o^* \in S_U$ and $y_o^* \in S_V$ with respect to the norms in X^* and Y^*, respectively. This implies that $T_o = x_o^* \otimes y_o^*$.

Now, suppose that $x_o^* \notin U_H^*$-dent B_U. Then we have:

There exists $\epsilon > 0$ such that for all $0 < \alpha \leq \epsilon$ and all

(2) $x^{**} \in S_{U_H^*}$ with $x^{**}(x_o^*) > 1 - \alpha$ there exists $x^* \in B_U \cap (x^{**} > 1 - \alpha)$ such that $\|x^* - x_o^*\| > \varepsilon$.

For $n \in \mathbb{N}$ with $n > 1/\varepsilon$ choose $h_n = h(1/n) \in S_H$ according to (1), and let $\gamma_n = \sup_{x^* \in B_U} |h_n(x^*,y_o^*)|$. Then we have: $\gamma_n = \sup_{x^* \in B_U} |h_n(x^*,y_o^*)| \geq h_n(x_o^*,y_o^*) > 1 - \alpha(1/n)$, so that, for $n > (1/\varepsilon)$ large enough: $(1 - \alpha(1/n))(1/\gamma_n) = 1 - \beta_n$ for some $0 < \beta_n \leq \varepsilon$. Let $x_n^{**} = (1/\gamma_n)h_n(\cdot,y_o^*)$. Then $x_n^{**} \in S_{U_H^*}$, and $x_n^{**}(x_o^*) > (1 - \alpha(\frac{1}{n}))/\gamma_n$, so that, according to (2), for all $n > (1/\varepsilon)$ large enough,

(3) there exist $x_n^* \in B_U \cap (x_n^{**} > (1 - \alpha(\frac{1}{n}))/\gamma_n)$ such that $\|x_n^* - x_o^*\| > \varepsilon$.

But we also have: $h_n(x_n^*,y_o^*) = \gamma_n x_n^{**}(x_n^*) > 1 - \alpha(\frac{1}{n})$, so that, according to (1): $\|x_n^* \otimes y_o^* - x_o^* \otimes y_o^*\|_{H^*} \leq \frac{1}{n} < \varepsilon$. We conclude that

$\varepsilon > \sup_{\substack{x \in B_X \\ y \in B_Y}} |(x_n^* - x_o^*)(x)y_o^*(y)| = \|x_n^* - x_o^*\|$, which contradicts (3). This shows that $x_o^* \in U_H^*$-dent B_U. Analogous arguments reveal that $y_o^* \in V_H^*$-dent B_V. This completes the proof.

We now use Propositions 1 and 2 to derive our results on weak*-denting points in duals of operator spaces.

3. <u>Theorem</u>: Let H be a linear subspace of $L_{w*}(X^*,Y)$, containing $X \otimes Y$: $X \otimes Y \subset H \subset L_{w*}(X^*,Y)$. Then we have: $w^*\text{-dent } B_{H^*} = w^*\text{-dent } B_{X^*} \otimes w^*\text{-dent } B_{Y^*}$. In particular:

$$w^*\text{-dent } B_{(X \tilde{\otimes}_\varepsilon Y)^*} = w^*\text{-dent } B_{(K_{w*}(X^*,Y))^*} = w^*\text{-dent } B_{(L_{w*}(X^*,Y))^*} =$$

$$= w^*\text{-dent } B_{X^*} \otimes w^*\text{-dent } B_{Y^*}.$$

<u>Proof</u>: We first note that $L_{w*}(X^*,Y) = B(X^*_\tau, Y^*_\tau)$, where, for a Banach space Z, we denote by Z^*_τ the dual of Z, endowed with the Mackey topology of the dual pair (Z^*,Z), i.e. the topology on Z^* of uniform convergence on the weakly compact disks in Z. (The fact that $L_{w*}(X^*,Y) = B(X^*_\tau,Y^*_\tau)$ can be read from Proposition 2.1 of [7].) Then we apply Propositions 1 and 2 for the situation $X \otimes Y \subset H \subset L_{w*}(X^*,Y) = B(X^*_\tau,Y^*_\tau)$, and note that, in this case, $X^*_H \subset X$ and $Y^*_H \subset Y$.

4. <u>Theorem</u>: Let H be a linear subspace of $L(X,Y)$ containing $X^* \otimes Y$: $X^* \otimes Y \subset H \subset L(X,Y)$. Then $w^*\text{-dent } B_{H^*} = \text{dent } B_X \otimes w^*\text{-dent } B_{Y^*}$. In particular:

$$w^*\text{-dent } B_{K(X,Y)^*} = w^*\text{-dent } B_{W(X,Y)^*} = w^*\text{-dent } B_{L(X,Y)^*} =$$

$$= \text{dent } B_X \otimes w^*\text{-dent } B_{Y^*}.$$

<u>Proof</u>: We apply Propositions 1 and 2 for the case $X^* \otimes Y \subset H \subset L(X,Y) \subset B(X,Y^*)$ with $U = X \subset X^{**}$ and $V = Y^*$, and note that, in this case, $U^*_H \subset X^*$ and $V^*_H \subset Y$.

5. <u>Theorem</u>: $\text{dent } B_{X \tilde{\otimes}_\pi Y} = \text{dent } B_X \otimes \text{dent } B_Y$.

<u>Proof</u>: We apply Propositions 1 and 2 for the situation $X^* \otimes Y^* \subset B(X,Y)$ with $U = X \subset X^{**}$ and $V = Y \subset Y^{**}$: If T_0 is a denting point of $B_{X \tilde{\otimes}_\pi Y}$, then $T_0 \in w^*\text{-dent } B_{B(X,Y)^*}$, and thus $T_0 = x_0 \otimes y_0$ with $(x_0,y_0) \in \text{dent } B_X \times \text{dent } B_Y$. Conversely, if $(x_0,y_0) \in \text{dent } B_X \times \text{dent } B_Y$ then $x_0 \in X \cap w^*\text{-dent } B_{X^{**}}$ and $y_0 \in Y \cap w^*\text{-dent } B_{Y^{**}}$, so that $T_0 = x_0 \otimes y_0 \in w^*\text{-dent } B_{B(X,Y)^*} = \text{dent } B_{X \tilde{\otimes}_\pi Y}$.

For more general operator spaces, we can deduce the following weaker results.

6. Corollary:

(a) If H is a linear subspace of $L(X^*,Y^{**})$ containing $X \otimes Y$: $X \otimes Y \subset H \subset L(X^*,Y^{**})$, then $w^*\text{-dent } B_{X^*} \otimes w^*\text{-dent } B_{Y^*} \subset w^*\text{-dent } B_{H^*} \subset \text{dent } B_{X^*} \otimes \text{dent } B_{Y^*}$.

(b) If H is a linear subspace of $L(X^{**},Y^{**})$ containing $X^* \otimes Y$: $X^* \otimes Y \subset H \subset L(X^{**},Y^{**})$, then $\text{dent } B_X \otimes w^*\text{-dent } B_{Y^*} \subset w^*\text{-dent } B_{H^*} \subset \text{dent } B_{X^{**}} \otimes \text{dent } B_{Y^*}$.

A particular consequence of our results on weak*-denting points in duals of operator spaces is the fact that, for $\dim X \geq 2$ and $\dim Y \geq 2$, none of the operator spaces $X \tilde{\otimes}_\varepsilon Y$, $K_{w^*}(X^*,Y)$, and $K(X,Y)$ (and thus none of their superspaces) has Mazur's intersection property. Recall that a Banach space Z is said to have Mazur's intersection property if every closed bounded convex subset of Z is the intersection of the closed balls containing it, and that this is the case if and only if the set of w*-denting points of B_{Z^*} is norm-dense in the unit sphere S_{Z^*} ([4,Thm. 2.1]).

7. Proposition: Assume that $\dim X \geq 2$ and $\dim Y \geq 2$. Then neither $X \tilde{\otimes}_\varepsilon Y$ nor $K(X,Y)$ has Mazur's intersection property.

Proof: We show that, for $\dim X \geq 2$ and $\dim Y \geq 2$, the dual of any linear space H with $X \otimes Y \subset H \subset K_{w^*}(X^*,Y)$ does not have its weak*-denting points norm-dense in its unit sphere. For, assume that $\| \,\|\text{-cl}(w^*\text{-dent } B_{H^*}) = S_{H^*}$. Then, according to Theorem 3, we have: $S_{H^*} = \| \,\|\text{-cl}(w^*\text{-dent } B_{X^*} \otimes w^*\text{-dent } B_{Y^*})$. According to Lemma 1.2 in section 1 of our paper [10], this implies that $S_{H^*} = S_{X^*} \otimes S_{Y^*}$, which contradicts our assumption that $\dim X \geq 2$ and $\dim Y \geq 2$.

In closing this paper, we should like to point out that, for general denting points in duals of operator spaces, the following result holds:

If H is a linear subspace of $K_{w^*}(X^*,Y)$ containing $X \otimes Y$: $X \otimes Y \subset H \subset K_{w^*}(X^*,Y)$, then $\text{dent } B_{H^*} = \text{dent } B_{X^*} \otimes \text{dent } B_{Y^*}$.

The proof of this result requires a somewhat more delicate analysis of (an extension of) the barycentric calculus for not necessarily continuous functions, and will be given in our paper [10].

REFERENCES

[1] Diestel, J.: Geometry of Banach Spaces - Selected Topics, Lecture Notes Math., Vol. 485; Springer, Berlin, Hdgb., New York, 1975.

[2] Diestel, J., Uhl, J.J.Jr.: Vector Measures, Amer. Math. Soc. Surveys 15, (1977).

[3] Giles, J.R.: Convex Analysis with Application in the Differentiation of Convex Functions, Pitman Adv. Publ. Program; Boston, London, Melbourne, 1982.

[4] Giles, J.R., Gregory, D.A., and Sims, B.: Characterization of normed linear spaces with Mazur's intersection property, Bull. Austr Math. Soc. 18(1978), 105-123.

[5] Grothendieck, A.: Produits tensoriels topologiques et espaces nucléaires, Amer. Math. Soc. Memoirs 16(1955).

[6] Phelps, R.R.: Dentability and Extreme Points in Banach Spaces, J. Funct. Analysis 16(1974), 78-90.

[7] Ruess, W.M.: [Weakly] Compact Operators and DF Spaces, Pacific J. Math. 98(1982), 419-441.

[8] Ruess, W.M.: Duality and Geometry of Spaces of Compact Operators, Proc. Paderborn Conf. Funct. Analysis: Surveys and Recent Results III. Eds.: K.-D. Bierstedt and B. Fuchssteiner. North-Holland Math. Studies 90(1984), 59-78.

[9] Ruess, W.M., Stegall, C.P.: Extreme Points in Duals of Operator Spaces, Math. Ann. 261(1982), 535-546.

[10] Ruess, W.M., Stegall, C.P.: Exposed and Denting Points in Duals of Operator Spaces, to appear in Israel J. Math.

[11] Smul'yan, V.L.: On some geometrical properties of the unit sphere in the space of the type (B), Math. Sbornik N.S. 6(48) (1939), 77-94.

[12] Smul'yan, V.L.: Sur la dérivabilité de la norme dans l'espace de Banach. Doklady Akad. Nauk SSSR (N.S.) 27(1940), 643-648.

SOME REMARKS CONCERNING THE KREIN-MILMAN AND THE RADON-NIKODYM PROPERTY OF BANACH SPACES

W. Schachermayer
Department of Mathematics
Universitat Linz
A-4040 Linz, Austria

Abstract:

We present an example, simplifying an earlier one due to R. C. James, of a 1-separated tree with empty wedge intersections such that its closed convex hull has continuum many extreme points.

1. Introduction: The present paper deals with the (still open) question, whether the two properties of Banach spaces mentioned in the title (and abbreviated RNP and KMP) are equivalent. It is based on [S] and we have tried to emphazise the importance of the notion of a "complemented bush". This concept was defined and studied in [H] and implicitly used in [S].

We give a simplified version of an example due to R. C. James [J]. In the language of trees and bushes we construct a tree such that its closed convex hull - as well as the closed convex hull of many of its asymptotic subtrees - has continuum many extreme points.

2. Definitions and notations: There are different ways of looking at the aspects relevant for RNP in Banach spaces: One may formulate these in terms of operators from L^1 to X or in terms of trees and bushes (which again are just special martingales). These are only different sides of the same coin. As we shall use the identification of a tree or a bush and an operator from L^1 to X often throughout this paper we shall recall this here in some detail.

Δ will denote the Cantor set $\{-1,+1\}^{\mathbb{N}}$ and m the normalized Haar-measure on Δ. The Rademacher-functions r_n on Δ are the projections onto the n'th coordinate. $\mathcal{B}$ will denote the Borel-σ-algebra of Δ and $\mathcal{B}_n$ the sub-σ-algebra generated by $r_1,\dots,r_n$.

We define a tree in a Banach space X to be a collection

$$T = \{x_{n,i} : 1 \leqq i \leqq 2^{n-1}, n \in \mathbb{N}\}$$

in the unit-ball of X such that for each (n,i)

$$x_{n,i} = 1/2(x_{n+1,2i-1} + x_{n+1,2i})$$

and such that there is a positive separation constant $\varepsilon > 0$ such that for each (n,i)

$$\| x_{n,i} - x_{n+1,2i-1} \| \geqq \varepsilon.$$

To a tree $\mathcal{T}$ we associate an operator T from $L^1(\Delta,m)$ to X in the following way: For (n,i), $n \in \mathbb{N}$, $1 \leqq i \leqq 2^{n-1}$ let $(\delta_1,\ldots,\delta_{n-1})$ be the unique element of $\{-1,1\}^{n-1}$ such that

$$\sum_{k=1}^{n-1} ((\delta_k+1)/2)\, 2^{n-k-1} = i-1$$

and let $I_{n,i}$ be the atom of $\mathcal{B}_n$ consisting of all $(\varepsilon_k)_{k=1}^{\infty}$ such that $\varepsilon_k = \delta_k$ for $k = 1,\ldots,n-1$. Define

$$T(2^{n-1} \chi_{I_{n,i}}) = x_{n,i}.$$

Then T extends by linearity and continuity to a bounded operator from $L^1(\Delta,m)$ to X.

We have treated the case of trees instead of the more general case of bushes (see [H] for a definition) for notational convenience only; but the above identification carries over to this case analogously except that we have to use in an obvious way a more tedious index set, a different Cantor-set and bigger finite σ-algebras Σ_n.

Hence a tree $\mathcal{T}$ gives an operator T from $L^1(\Delta,m)$ to X which is easily (almost by definition) seen to be not representable. Conversely, given an operator from $L^1(\Omega,\Sigma,\mu)$ to X which is not representable one may find a subset $A \subseteq \Omega$, $\mu(A) > 0$ and an increasing sequence of finite σ-algebras Σ_n such that by the identification sketched above the traces of Σ_n on A define a bush with strictly positive separation constant (see [ST] for a detailed exposition).

In the sequel we restrict us to the notationally more convenient case of trees and operators from $L^1(\Delta,m)$ corresponding to a tree.

There is a natural ordering among the indices (n,i) of a tree, which is obtained from $(n+1,j) > (n,i)$ if $j = 2i-1$ or $j = 2i$. A <u>branch</u> of T is a sequence $(x_{n,i_n})_{n=1}^{\infty}$ such that $((n,i_n))_{n=1}^{\infty}$ is totally ordered. A <u>wedge</u> [H] of the tree T is a set of the type

$$W_{n,i} = \{x_{(m,j)} : (m,j) > (n,i)\}.$$

Recall the following notation from [S] and [ST] for an operator $T : L^1(\Delta,m) \to X$ and $A \in \mathcal{B}$, $m(A) > 0$:

$$L_A = \{Tf : f = f\chi_A \geq 0 \text{ and } \int f dm = 1\}.$$

The connection between wedges of a tree T and these sets L_A for the corresponding operator T is given - in the case where A equals an atom $I_{n,i}$ - by the easily verified formula

$$\overline{L}_{I_{n,i}} = \overline{\text{co}}\, W_{n,i}$$

where $\overline{\text{co}}$ denotes the closed convex hull.

<u>2.1. Definition:</u> a) We say that a tree has empty wedge <u>intersections</u> if for every branch $((n,i_n))_{n=1}^{\infty}$

$$\bigcap_{n=1}^{\infty} \overline{\text{co}}\, W_{n,i_n} = \emptyset$$

b) We say that a tree is <u>complemented</u> ([H]) if there is $\Theta > 0$ such that for each (n,i)

$$\| u-v \| \geq \Theta \min (\|u\|, \|v\|)$$

for every $u \in$ linear span $(W_{n+1,2i-1})$ and $v \in$ linear span $(W_{n+1,2i})$.

It was shown in [H] and [S] that for a tree T satisfying a) and b)

$$\overline{\text{co}}\,(T)$$

has no extreme points. In fact, it was noticed in [S] that one may even allow $\Theta > 0$ to depend on (n,i) and still arrive at the same conclusion.

It was shown in [ST] that, starting from any non-representable operator $T : L^1(\Sigma,\Omega,\mu) \to X$, one may associate (as sketched above) a bush, which even has empty wedge intersections (with the obvious extension of the above definition to bushes). Hence the problem of showing that KMP implies RNP is solved if we can achieve, in addition, the complementation property of b).

The aim of section 3 of the present paper is to give an example of a "badly uncomplemented tree" with empty wedge intersections but such that

$$\overline{\text{co}}\ (\mathcal{T})$$

(and in fact $\overline{L}_A$ for each $A \in \Sigma$, $m(A) > 0$) has continuum many extreme points and equals the convex hull of its extreme points.

<u>3. A tree whose closed convex hull has many extreme points:</u> We define an operator $T : L^1(\Delta) \to c_0$ coordinatewise. For the odd coordinates it will be the well-known "Rademacher operator" which is one of the arch-examples of a non-RNP-operator:

$$(Tf)_{2n-1} = \langle f, r_n \rangle \qquad n = 1,2,\ldots$$

On the even coordinates we want T to be a compact operator, which is chosen in such a way that $T^* : \ell^1 \to L^\infty(\Delta)$ maps ℓ^1 into a dense subset of $C(\Delta)$. E.g., let $(x_n)_{n=1}^\infty$ be a dense sequence in the unit-ball of $C(\Delta)$ and let

$$(Tf)_{2n} = 2^{-n} \langle f, x_n \rangle \qquad n = 1,2,\ldots$$

Clearly T is not a RNP-operator.

In fact the tree $\mathcal{T}$ associated to T has separation constant 1 and empty wedge intersections: Indeed for each (n,i) each element of

$$\overline{\text{co}}\ (W_{n,i})$$

has entries of absolute value 1 in the coordinates $1,3,\ldots,2n-1$.

Proposition 3.1: For any measurable set $A \subseteq \Delta$, $m(A) > 0$, the set

$$K_A = \overline{L_A}$$

has continuum many extreme points and equals the closed convex hull of its extreme points.

Proof: Note that

$$T^* : \ell^1 \to L^\infty(\Delta)$$

takes its values in $C(\Delta)$. Hence we may define (with an abuse of notation)

$$T^{**} : M(\Delta) \to \ell^\infty,$$

which is one to one by the density of the range of T^* in $C(\Delta)$.

Now let us first assume that $A = \Delta$. Let

$$\mathcal{K}_\Delta = \{T^{**}(\mu) : \mu \text{ probability measure on } \Delta\}$$

$\mathcal{K}_\Delta$ is a σ^*-compact subset of ℓ^∞ and it is easily verified that

$$K_\Delta = \mathcal{K}_\Delta \cap c_o.$$

Now let $t = (\varepsilon_k)_{k=1}^\infty$ and $t' = (\varepsilon_k')_{k=1}^\infty$ be elements of Δ satisfying

$$\varepsilon_k = -\varepsilon_k' \text{ for all but finitely many } k. \qquad (*)$$

We claim that

$$e(t,t') = T^{**}(1/2(\delta_t+\delta_{t'}))$$

is an extreme point of K_Δ, where δ_t denotes the Dirac-measure at t.

Indeed $e(t,t')$ belongs to c_o because the odd coordinates are eventually zero while the even coordinates tend to zero (without any problem). On the other hand the segment

$$E(t,t') = \{\lambda\delta_t+(1-\lambda)\delta_{t'} : 0 \leqq \lambda \leqq 1\}$$

is an extremal set in the simplex of probability measures on Δ. From the injectivity of T^{**} we get that $T^{**}(E(t,t'))$ is an extremal set in K_Δ. As $T^{**}(\delta_t)$ evidently is not in K_Δ we infer that $e(t,t')$ is the only point of $T^{**}(E(t,t')) \cap K_\Delta$ and therefore an extreme point of K_Δ.

Let us finally show that K_Δ is the closed convex hull of the extreme points $e(t,t')$. Indeed let $s = (\delta_k)_{k=1}^{\infty}$ in Δ be given and define $s^n = (\delta_k^n)_{k=1}^{\infty}$ by

$$\delta_k^n = \delta_k \qquad k = 1,\dots,n$$

$$\delta_k^n = -\delta_k \qquad k > n$$

Clearly the pairs (s,s^n) satisfy (*) and

$$\lim_{n\to\infty} (1/2(\delta_s + \delta_{s^n})) = \delta_s$$

the limit taken in the σ^*-topology of $M(\Delta)$. Hence the σ^*-closed convex hull of all $(1/2(\delta_t + \delta_{t'}))$ with (t,t') satisfying (*) are all the probability measures on Δ. It follows from the $\sigma(M(\Delta), C(\Delta)) - \sigma(\ell^\infty, \ell^1)$ - continuity that K_Δ is the σ^*-clossed convex hull of the corresponding $e(t,t')$, hence K_Δ the σ-closed (and therefore norm-closed) convex hull of the extreme points $e(t,t')$.

Let us now pass to the case of $A \in \mathcal{B}$, $m(A) > 0$ instead of the whole set Δ. Let $\tilde{A}$ be the closure of all points of Lebesgue-density one of A. It follows easily that

$$K_{\tilde{A}} := \{T^{**}\mu : \mu \text{ probability measure supported by } \tilde{A}\}$$

equals the σ^*-closure of

$$K_A := \bar{L}_A.$$

Again the points $e(t,t')$ such that $t,t' \in \tilde{A}$ and t,t' satisfy (*) are extreme points of K_A and an argument similar to the above and using the Lebesgue density shows that the closed convex hull of these $e(t,t')$ equals K_A.

□

Remarks: (1) The example is astonishingly simple; it consists only of a compact perturbation of the well-known "Rademacher-operator".

However, it is entirely based on the previous very technical one of James [J], which displays essentially the same phenomena.

It was only by gradually understanding and simplifying James' example that the author finally arrived at this version.

Our example is a tree-version (this corresponds to $c_i = 2$ for $i \geqq 1$ in [J], which implies that the function g is just constant on Δ_{11}) of James' example, the functions $\Phi_{n,k}$ in [J] correspond to the "Rademacher-part" of our T while the functions ω_p correspond to the "compact part" of T.

Finally let us illustrate why the tree $\mathcal{T}$ corresponding to our T is "badly uncomplemented" i.e. badly fails condition b) of definition 2.1.

Of course we know from [H] and [S] that $\mathcal{T}$ cannot be complemented (otherwise K_Δ could not have extreme points). But it seems worth while to see this explicitly.

Fix any tree index (n,i). We shall show that the linear spans of the wedges $W_{n+1,2i-1}$ and $W_{n+1,2i}$ are not complemented: Let $m > n+1$ and find j such that $I_{m,2j-1} \subseteq I_{n+1,2i-1}$. Then

$$x_{m,2j-1} - x_{m,2j} \in \text{lin sp } W_{n+1,2i-1}$$

and

$$\|x_{m,2j-1} - x_{m,2j}\| = 2^{m-1} \|T(\chi_{I_{m;2j-1}} - \chi_{I_{m,2j}})\| \geqq$$

$$\geqq 2^{m-1} |\langle \chi_{I_{m,2j-1}} - \chi_{I_{m,2j}}, r_m\rangle| = 2$$

Similarly,

$$x_{m,2j+2^{m-n-1}-1} - x_{m,2j+2^{m-n-1}} \in \text{lin sp } W_{n+1,2i}$$

and

$$\| x_{m,2j+2^{m-n-1}-1} - x_{m,2j+2^{m-n-1}} \| \geqq 2.$$

On the other hand

$$\lim \| x_{m,2j-1} - x_{m,2j} - (x_{m,2j+2^{m-n-1}-1} - x_{m,2j+2^{m-n-1}}) \| = 0.$$

Indeed, all the odd coefficients of the above element of c_o are zero while the even coordinates tend to zero uniformly. This shows that the tree T is badly uncomplemented.

Finally let us observe why the present example cannot furnish a counter-example to the general questions of whether RNP and KMP are equivalent: If X denotes the closure of the space spanned by T then X contains a subspace isomorphic to c_o (as does any infinite-dimensional subspace of c_o). Hence X has no chance to have KMP.

However, this shows one interesting fact: So far, all the pathologies arising in the absence of RNP could be shown to happen in some set of the form K_A (see [ST] for a convincing presentation of this fact). But the present example shows that if one tries to prove the equivalence of RNP and KMP one may not restrict oneself to such sets but has to adopt other methods of constructing "bad" sets.

REFERENCES

[B-T] J. Bourgain, M. Talagrand: Dans un espace de Banach réticulé solide, la propriété de Radon-Kikodym et celle de Krein-Milman sont equivalent. Proc. AMS 81 (1981), p. 93-96.

[H] A. Ho: The Krein-Milman property and complemented bushes in Banach space, Pac. J. of Math. 98 (1982), p. 347-363.

[J] R.C. James: Extreme points and an unusual Banach space, appeared in Banach space theory and its applications, Springer lecture notes 991 (1983), p. 111-123.

[J-T] J. Lindenstrauss, L. Tzafriri: Classical Banach spaces, vol. 1, Springer 1977.

[S] W. Schachermayer: For a Banach space isomorphic to its square the Radon-Nikodym property and the Krein-Milman property are equivalent, to appear in Studia Math. (1985).

[ST] C. Stegall: The Radon-Nikodym property in Banach spaces, Part I, appeared in "Vorlesungen uas dem FB. Mathematik der Universität Essen, Heft 10, Essen (1983).

ON THE NORMS OF SOME PROJECTIONS*

Boris Shekhtman
Department of Mathematics
University of California
Riverside, CA 92521

I. Introduction

In this note we summarize some results about the norms of the projections onto subspaces that are natural in Approximation Theory. We are mostly concerned with the subspaces in $C(K)$ spaces although most of the results are also valid for $L_1(K)$ spaces.

We use the usual notations: $\lambda(X)$ for the projectional constant of X and $d(X,Y)$ for the Banach-Mazur distance.

II. Preliminary Results

In this section we collect some technical results that we list as propositions.

Proposition 1 (Olevskii [4]). Let (K,μ) be a probability space and $(\phi j)_{j=1}^{\infty}$ be an orthonormal system in $L_2(\mu)$. Suppose in addition $\|\phi_j\|_\infty = O(1)$. Then for the projection F_n given by

$$F_n f = \sum_{j=1}^{n}\left(\int \phi_j \cdot f d\mu\right)\phi_j$$

acting from $L_\infty(\mu) \to L_\infty(\mu)$, we have $\limsup \frac{\|F_n\|}{\log n} > 0$.

Proposition 2. Let $(\phi_j)_{j=1}^{n}$ be a sequence of functions from $L_\infty(\mu)$ where μ is a probability measure. Let $(\eta_j)_{j=1}^{n}$ be a sequence of positive numbers and let f be a norm one positive functional on $L_\infty(\mu)$ such that $f(|\Sigma c_j\phi_j|) \geq \Sigma\, \eta_j |c_j|$ for any set of real numbers (c_j). Consider an operator $A_n : L_\infty(\mu) \to L_\infty(\mu)$ given by $A_n = \Sigma \mu_j \otimes \phi_j$ where $\mu_j \in [L_\infty(\mu)]^*$. Then $\|A_n\| \geq \Sigma \eta_j \|\phi_j\|$

Proof: Let σ be a probability measure corresponding to f. Let

*This research was supported by National Science Foundation Grant MCS-8301646.

$\nu = \Sigma|\mu_j|/2^j$. Then there are functions $g_j \in L_1(\nu)$ so that

$$A_n x = \int (\Sigma g_j(s)\phi_j) x(s) d\nu$$

and

$$\|A_n\| = \mu - \operatorname{ess\,sup}_t \int |\Sigma g_j(s)\phi_j(t)| d\nu(s)$$

$$\geq \int\int |\Sigma g_j(s)\phi_j(t)| d\nu(s) d\sigma(t)$$

$$= \int [\int |\Sigma g_j(s)\phi_j(t)| d\sigma(t)] d\nu(s)$$

$$\geq \int (\Sigma \eta_j |g_j(s)|) d\nu(s)$$

$$= \Sigma \eta_j \|\mu_j\|.$$

Proposition 3. Let G be a compact connected group. Then there exists a universal constant C such that for $(\phi_j)_{j=1}^n$ an increasing (with respect to an order on $\hat{G}$) sequence of cotinuous characters on G,

$$\int |\sum_{j=1}^n c_j \phi_j| d\mu \geq C \sum_{j=1}^n \frac{1}{j}|c_j|$$

for any sequence (c_j).

Proof: The proof is the same as in the case $G = T$. It consists of the construction of the function $F \in C(G)$ so that $\|F\|_\infty \leq 1$, $\int \bar{F}\phi d\mu = 0$ for $\phi > \phi_n$ and $\operatorname{Re}[c_j \cdot \int \bar{F}\phi_j d\mu] \geq C\frac{|c_j|}{j}$. As in [1] we represent $(\phi_j) = \cup(\phi_j)_{j=n_k}^{n_{k+1}}$ so that $n_{k+1} - n_k = 4^k$ and construct $F_k = \frac{1}{4^k}\sum_{j=n_k}^{n_{k+1}} \phi_j \cdot d_j$ where $d_j = \frac{c_j}{|c_j|}$. Let h_k be a harmonic conjugate (with respect to the linear order in $\hat{G}$ [5]) to the function $\frac{1}{4}|F_k|$. We define $G_0 = F_0$, $G_{j+1} = G_j e^{-h_{j+1}} + F_{j+1}$ and $F = G_n$. The proof that the function F has the desired properties is identical with the one in [1].

III. Projections Onto Translation-invariant Subspaces

In this section we give some immediate corollaries from Propositions 1-3.

Theorem 1. Let G be a compact connected group. Then there exists a constant C so that for $(\phi_j)_{j=1}^n$ a set of continuous characters on G,

$$\lambda([\phi_j]_{j=1}) \geq C \cdot \log n .$$

Proof: The combination of Propositions 2 and 3 is the proof.

Remark 1. It follows immediately from this theorem that

$$\lambda([\phi_j]_{j=1}^n) \leq d([\phi_j],\ell_\infty^{(n)}) \leq e^{\lambda([\phi_j])} .$$

Now consider G to be the unit circle and $\phi_j = z^{\lambda_j}$, where (λ_j) is a sequence of integers. It is well known that if $\lambda_j = j$, then $\lambda([\phi_j]_{j=1}^n) \sim \log n$. If (λ_j) is a lacunary sequence then $([\phi_j]_{j=1}^n) \sim \sqrt{n}$. In the latter case, $d([\phi_j],\ell_\infty^{(n)}) \sim \sqrt{n}$ also. A similar estimate can be done for $\lambda_j = j$.

Theorem 2. $d([z^j]_{j=1}^n, \ell_\infty^{(n)}) \sim \lambda\, [z^j]_{j=1}^n \sim \log n$.

Proof: We only have to prove that $d([z^j], \ell_\infty^{(n)}) \leq C \cdot \log n$. Let $P_n = \Sigma\, \delta_{z_j} \otimes \phi_j$ be an interpolating projection onto $[z^j]$ that interpolates on the roots of unity. Then for any sequence of numbers a_j, we have

$$\max|a_j| = \max_j |\delta_{z_j}(\Sigma\, a_j\phi_j)| \leq \|\Sigma a_j \phi_j\| .$$

On the other hand let f be a function of norm $\max|a_j|$ so that $f(z_j) = a_j$, then

$$\|\Sigma a_j \phi_j\| = \|P_n f\| \leq \|P_n\| \max|a_j| ,$$

but the norm of $\|P_n\|$ is known to grow as $\log n$ (cf. [3]).

We now turn our attention to the more general groups.

Theorem 3. Let G be a compact group. Let $(\phi_j)_1^n \subset \hat{G}$. Let $F_n = \sum_1^n \phi_j \otimes \phi_j$. Then $\|F_n\| = \int |\Sigma\phi_j| d\mu = \lambda\,[\phi_j]_{j=1}^n)$.

Proof: Following the standard argument, let P_n be an arbitrary projection from $C(G)$ onto $[\phi_j]$. For every $s \in G$ define $T_s : C(G) \to C(G)$ by $(T_s x)(t) = x(t \cdot s)$, $t \in G$. Then it is easy to check that

$$(F_n f)(t) = \int (T_{s^{-1}} P_n T_s f) d\mu_s$$

and hence $\|P_n\| \geq \|F_n\|$.

Theorem 4. Let G be a compact group such that the identity component of G has finite index m. Then there exists a constant C_m depending on m only, such that

$$\lambda([\phi_j]_{j=1}^n) \geq C_m \cdot \log n .$$

Proof: It allows that the dual group has exactly m elements of finite order. Now factoring it out and using Proposition 3, we get

$$\int |\sum_{j=1}^n \phi_j| d\mu \geq C_m \cdot \log n \qquad (n > m).$$

If $\hat{G}$ has an infinite torsion subgroup, then Theorem 4 does not hold.

Theorem 5. Let $\Gamma \subset \hat{G}$ be a finite subgroup of the dual of an arbitrary compact group G. Then

$$\lambda([\phi]_{\phi\in\Gamma}) = d([\phi], \ell_\infty^{(\#\Gamma)}) = 1.$$

Proof: For every $\phi_0 \in \Gamma$ we have $\sum_{\phi\in\Gamma} \phi = \phi_0 \cdot \sum_{\phi\in\Gamma} \phi$. Hence for every point $t \in G$ we have either $\sum_{\phi\in\Gamma} \phi(t) = 0$ or $\phi(t) = 1$ for all $\phi \in \Gamma$. One way or the other, $\sum_{\phi\in\Gamma} \phi \geq 0$ and using Theorem 3, we get $\lambda([\phi]) = \int (\sum_{\phi\in\Gamma} \phi) d\mu = 1$.

This situation comes up when considering the Walsh functions or more generally, any periodic multiplicative system. Let $(\phi_j)_{j=1}^{\infty}$ be such a system. Then (ϕ_j) can be modeled as the set of characters of some compact group. The group generated by (ϕ_j) has infinite torsion subgroups $\phi_j^2 = 1$. By theorem 3 and Proposition 1 we conclude that $\limsup \lambda([\phi_j]_{j=1}^{n}) \to \infty$; yet, Theorem 5 shows that the "lim sup" in this statement cannot be replaced by "lim".

IV. Projections Onto Algebraic Polynomials

In this section we turn our attention to the subspaces of algebraic polynomials on the real interval [a,b].

Theorem 6. There exists a universal constant C such that for any increasing set of integers $(\lambda_j)_{j=1}^{n}$,

$$\lambda([\cos\lambda_j\theta]_{j=1}^{n}) \geq C \cdot \log n$$

$$\lambda(\sin\lambda_j\theta\ _{j=1}^{n}) \geq C \cdot \log n .$$

Proof: We would like to estimate the sum $\int_{-\pi}^{\pi} | \Sigma a_j \cos\lambda_j\theta| d\theta$. Using $\cos\lambda_j\theta = (e^{i\lambda_j\theta} + e^{-i\lambda_j\theta})/2$ and Proposition 3, we easily get a constant C such that

$$\int_{\pi}^{\pi} | \sum_{j=1}^{n} a_j \cos\lambda_j\theta| d\theta \geq C \cdot \sum_{j=1}^{n} \frac{|a_j|}{n-j} .$$

For an arbitrary projection $P_n = \Sigma\ \mu_j \otimes \cos\lambda_j\theta$, we have $\mu_j(\cos\lambda_j\theta) = 1$ and hence $\|\mu_j\| \geq 1$. Now by Proposition 2 we obtain the first inequality. The second one can be obtained in the same way.

Corollary 1. Let T_j and U_j be Chebyshev polynomials of the first and second kind on an arbitrary interval [a,b]. Let $(\lambda_j)_{j=1}^{n}$ be a sequence of integers. Then

$$\lambda([T_{\lambda_j}]_{j=1}^{n}) \geq C \cdot \log n$$

$$\lambda([U_{\lambda_j}]_{j=1}^{n}) \geq C \cdot \log n$$

Proof: It follows from the fact that the distances $d([T_{\lambda_j}]^n,[\cos\lambda_j\theta]^n)$ and $d([U_{\lambda_j}]^n,[\sin\lambda_j\theta]^n)$ are bounded uniformly in n and (λ_j) (cf.[3]).

Corollary 2. Consider $[t^j]_{j=0}^n \subset C_{[a,b]}$. Then

$$\lambda([t^j]_{j=0}^n) \sim d([t^j]_{j=0}^n,\ell_\infty^{(n+1)}) \sim \log n .$$

Proof: Since $[t^j]_{j=0}^n = [T_j]_{j=0}^n$ we have $\lambda([t^j]) = \lambda([T_j]) \geq C \cdot \log n$ and hence $d([t^j],\ell_\infty^{(n+1)}) \geq C \log n$.

To prove the reverse inequality, we again consider the interpolating projection P_n from $C_{[a,b]} \to [t^j]_{j=0}^n$ that interpolates at the zeros of the T_{n+1}. The norm of P_n is known to grow as $\log n$. The rest of the argument is the same as in Theorem 2.

In view of Corollaries 1 and 2, one is tempted to ask:

Problem 1. Does Corollary 1 hold for other sequences of orthogonal polynomials?

Problem 2. Does Corollary 2 hold if we replace $[t^j]_{j=0}^n$ by $[t^{\lambda_j}]_{j=1}^n$ for some (λ_j)?

In the rest of the section we will describe some results about Problem 2. Proposition 4 is a quick consequence of Corollary 2.

Proposition 4. Let $\lambda_j \leq j + o(\log^2 j)$. Then $\lambda([t^{\lambda_j}]_{j=1}^n) \to \infty$ as $n \to \infty$.

Proof: Let $M = [t^j]_{j=1}^{\lambda_n}$ and P_n be a projection from $C_{[0,1]}$ onto $[t^{\lambda_j}]_{j=1}^n$. Then $M = [t^{\lambda_j}]_{j=1}^n \otimes N$, where $N = \ker(P_n|M)$; $\dim N \leq o(\log^2 n)$. Then there exists a projection Q_n from $C_{[0,1]}$ onto N with the norm $\|Q_n\| \leq o(\log n)$. It is easy to see that $P_n + Q_n - Q_nP_n$ is a projection onto M, and by Corollary 2 we have

$$\log\lambda_n \leq \|P_n + Q_n - Q_nP_n\| \leq (1 + \|P_n\|)(\|Q_n\| + 1).$$

Hence $1 + \|P_n\| \geq \frac{\log \lambda n}{\|Q_n\|} \to \infty$ as $n \to \infty$.

Next we consider one very special case.

Theorem: Let $\lambda_j = j^2$. Then $\lambda[t^{\lambda_j}]_{j=1}^n) C \cdot \log n$.

Proof: Without loss of generality we can assume $t = \cos\theta$. Then any projection P_n from $C_{[-\pi,\pi]}$ onto $[\cos^{\lambda_j}\theta]_{j=1}^n$ can be written as

$$P_n = \Sigma\, \mu_j \otimes \cos^{\lambda_j}\theta.$$

On the other hand, $\cos\theta = \frac{e^{i\theta} + e^{-1\theta}}{2}$. Hence

$$\cos^k\theta = \frac{1}{2^k}\,\Sigma \binom{k}{j} e^{i(k-2j)\theta}. \tag{1}$$

Therefore the projection P_n can be written as

$$P_n = \sum_{j=1}^{\lambda_n} \nu_j \otimes e^{i(\lambda n-2j)\theta},$$

where ν_j are some linear combinations of μ_j. By Propositions 2 and 3 we get $\|P_n\| \geq \Sigma^{\lambda_n} \frac{1}{j}\|\nu_j\|$, and the proof boils down to estimating the norms $\|\nu_j\|$.

For sufficiently large n and m, let

$$\frac{m^2}{2} - \frac{m}{2} \leq j \leq \frac{m^2}{2} + \frac{m}{2}. \tag{2}$$

At this point we have to remark that for sufficiently large k, the coefficients in the polynomial (1) are distributed almost normally and thus there are $\approx \sqrt{k}$ coefficients that are the size $\frac{1}{\sqrt{k}}$ (we will call them essential) and the rest are negligible.

Now we estimate

$$\mu_j(\cos^{m^2}\theta - \cos^{(m-1)^2}\theta)$$

Since P_n is a projection, this is the j-th coefficient in the polynomial $\cos^{m^2}\theta - \cos^{(m-1)^2}\theta$. By (2) this coefficient is $\sim \frac{1}{m}$ since the corresponding coefficient in $\cos^{(m-1)^2}\theta$ is negligible. Thus $\mu_j(\cos^{m^2}\theta - \cos^{(m-1)^2}\theta) = \frac{1}{m}$. On the other hand,

$$\|\cos^{m^2}\theta - \cos^{(m-1)^2}\theta\| \sim \frac{1}{m}$$

and $\|\mu_j\| \geq 1$. Hence

$$\sum_{(m^2-m)/2}^{(m^2+m)/2} \frac{1}{j} \|\mu\| \geq \sim \ln \frac{m+1}{m-1} = \ln\left(1 + \frac{2}{m-1}\right)^{\frac{m-1}{2} \cdot \frac{2}{m-1}}$$

$$= \frac{2}{m-1} \ln \left[1 + \frac{2}{m-1}\right]^{\frac{m-1}{2}} \sim \frac{2}{m-1}.$$

Thus $\|P_n\| \geq \sum_{m>N}^{n} \frac{2}{m-1} \sim \log n.$

Similarly to Proposition 4, we can obtain Proposition 5.

Proposition 5. Let $|\lambda_j - j^2| \leq o(\log^2 j)$. Then $\lambda([t^{\lambda_j}]_{j=1}^n) \to \infty$ as $n \to \infty$

In Section III we observed that the projectional constant of the complex polynomials is the largest when (λ_j) is lacunary. The algebraic polynomials exhibit the opposite behavior.

Proposition 6. Let (λ_j) be lacunary; the

$$\lambda([t^{\lambda_j}]_j^n = 1) \leq d([t^{\lambda_j}]_{j=1}^n, \ell_\infty^{(n)}) \leq O(1).$$

Proof: Our proof follows from the results of [2], where it is proved that (t^{λ_j}) spans a copy of c_0 and forms a basis in its span.

I am grateful to Professors W. Johnson and L. Harper for many useful suggestions.

REFERENCES

[1] McGehee, O. C., L. Pigno, and B. Smith, "Hardy's inequality and the L^1 norm of exponential sums," Annals of Math. 113(1981), 613-618.

[2] Gurarri, V.I., and V.I. Macaev, "Lacunary power series in the spaces C and L_p," Izv. Akad. Nauk SSSR Ser. Mat. 30(1966), 3-14.

[3] Natanson, I.P., Constructive Theory of Functions, Gostekizdat, Moscow, 1949.

[4] Olevskii, A.M., Fourier Series with Respect to General Orthonormal Systems, Ergebuisse der Math. No. 86, Springer-Verlag, Berlin, 1975.

[5] Rudin, W., Fourier Analysis on Groups, Interscience Tracts in Pure and Appl. Math. No 12, J. Wiley & Sons, New York, 1962.

MORE GATEAUX DIFFERENTIABILITY SPACES

C. Stegall
Department of Mathematics
Johannes Kepler University
A-4040 Linz, Austria

We expand and (perhaps) clarify the results of [S4] and [S5]. The results given here, combined with a result of Debs [D] (see also [S6]) yield:

<u>Theorem 5:</u> Let I be a set and $\{X_i : i \in I\}$ a family of weakly k-analytic Banach spaces; then

$$(\sum_i X_i)_{l_p} \qquad 1 < p < \infty$$

and

$$(\sum_i X_i)_{c_o}$$

are weak Asplund spaces. Indeed, they are hereditarily weak Asplund spaces.

We consider only Hausdorff spaces.

We define C to be the class of all topological spaces such that: $K \in C$ if and only if we have a "fibration" type diagram:

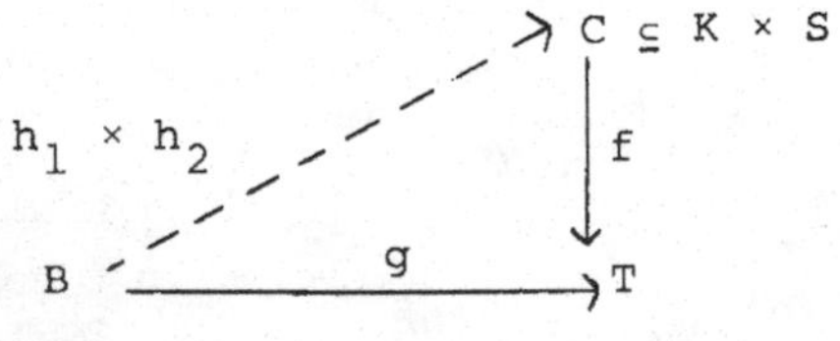

That is, if we are given topological spaces S,T,B,C with

$$C \subseteq K \times S,$$

$$B \text{ Baire},$$

and functions f and g with

$$f \text{ perfect, onto,}$$

$$g \text{ continuous,}$$

then there exist functions

$$h_1 : B \to K$$

$$h_2 : B \to S$$

so that $f \circ (h_1 \times h_2) = g$ and h_1 is continuous at each point of a dense G_δ subset of B (see [S4]).

The following is routine and is included as a mnemonic device to help the reader remember the basic properties of $\mathcal{C}$.

Lemma a: A topological space K is in $\mathcal{C}$ if and only if for every usc multivalued map Φ from the Baire space B to K such that $\Phi(b)$ is compact and non-empty there exists $\lambda : B \to K$ such that $\lambda(b) \in \Phi(b)$ and λ is continuous at each point of a dense G_δ set.

Proof: If $K \in \mathcal{C}$, we may take $S = T = B$, g = identity on B and let

$$C = \{(k,b) : k \in \Phi(b)\}.$$

It is routine, that

$$\mathrm{proj}_B : C \to B$$

is perfect and onto and the function $h_1 = \lambda$ has the desired properties (note: $h_2 = \mathrm{iden}_B$).

For the converse, if we have everything as in the diagram, define

$$\Phi(b) = \mathrm{proj}_K\, f^{-1}\, g(b).$$

Again, it is routine that Φ has all of the desired properties. Define $h_1 = \lambda$ and choose h_2 by the axiom of choice so that $(h_1 \times h_2)(b) \in C$ and $f(h_1 \times h_2)(b) = g(b)$.

The details of the following properties of C can be found in [S4], [S5]:

(C i) if $K_1 \in C$ and $f : K_1 \to K_2$ is perfect onto then $K_2 \in C$;

(C ii) if $K_1 \in C$ and $f : K_2 \to K_1$ is continuous and $1 : 1$, then $K_2 \in C$;

(C iii) C is closed under countable products;

(C iv) if $K = \bigcup_{n=1}^{\infty} K_n$ and each K_n is a closed subset of K and K_n is in C, then $K \in C$.

We define S to be the class of Banach spaces such that $X \in S$ if and only if $(X^*, \sigma(X^*,X))$ is in C.

The permanence properties of C translate into the following properties of S:

(S i) if $X \in S$ and $T : X \to Y$ is an operator with dense range then $Y \in S$;

(S ii) if X is in S and $T : Y \to X$ is an operator such that T^{**} is $1 : 1$, then Y is in S (in particular, if Y is a closed linear subspace of X, then $Y \in S$).

(S iii) if $\{X_n\}_{n=1}^{\infty} \subseteq S$ then $(\sum_n X_n)_{l_1} \in S$;

(S iv) if I is an arbitrary set and for each i, $X_i \in S$, then $(\sum_i X_i)_{c_0} \in S$ (and, from (S ii), $(\sum_i X_i)_{l_p} \in S$ for $1 < p < +\infty$)

We recall that the class S is non-trivial:

<u>Theorem A:</u> Let X be a Banach space. Then, each separable subspace of X has a separable dual if and only if for any bounded subset C

of X^*, any $\varepsilon > 0$, there exists an $x \in X$ such that

$$\operatorname{diam} \{x^* \in C : x^*(x) \geqq \sup_C x - \delta\} < \varepsilon.$$

One direction of the proof is found in [S1]; the other in [S2].

We shall call a Banach space X an Asplund space if X satisfies the conditions above.

A complete proof of the following can be found in [S4] (see also [S3]).

Theorem B: Let X be an Asplund space (respectively, $X \in S$). Let $g : X \to \mathbb{R}$ be a continuous convex function. Let U be an open subset of some Banach space Y and $f : U \to X$ continuous and Fréchet (resp. Gateaux) differentiable at each point of a dense G_δ subset of U. Then, $g \circ f$ is Fréchet (resp. Gateaux) differentiable at each point of a dense G_δ subset of U.

The proof of the Fréchet differentiable part is based on the well-known:

Lemma b: Let $\Phi : B \to P(K)$ be a minimal usc compact valued map. Fix $b_o \in B$, $k_o \in \Phi(b_o)$, $b_o \in U$ open, $k_o \in V$ open. Then, there exists a non-empty W open, $W \subseteq U$ such that $\Phi(W) \subseteq V$.

From this Lemma and Theorem A above we have that if B is an Asplund space and $\Phi : B \to P(X^*)$ as above, with the weak* topology on X^*, then for each $\varepsilon > 0$ and for each open $\emptyset \neq U \subseteq B^o$, there exists $\emptyset \neq V$ open, $V \subseteq U$, so that diam $\Phi(V) < \varepsilon$. (Diameter is in the norm on X^*.) This means of course that Asplund spaces $\subseteq S$. Our main result here is a "charting" lemma from which $(S\ ii)$, $(S\ iv)$ and other properties of S follow.

We need a few truly elementary results from topology.

Proposition A: Suppose $p : T \to B$ is a minimal perfect map onto the Baire space B and $g : T \to \mathbb{R}$ is lower (or upper) semicontinuous. Then, there exists a dense G_δ subset of G of B such that

(i) for all $b \in G$, g is constant on $p^{-1}(b)$;

(ii) there exists a function $h : B \to \mathbb{R}$ so that $b \in G$, $h(b) = g(t)$, $t \in \Phi(b)$ and h is continuous at each point of G.

Proof: We suppose that g is lsc (replace g by $1-g$ if g is usc). Define

$$T_{n,i} = \{t : g(t) \leq i/2^n\} \quad n = 1,2,\ldots \quad -\infty < i < +\infty .$$

Since p is perfect, $B_{n,i} = p(T_{n,i})$ is also closed and $\bigcup_i B_{n,i} = B$. Since B is Baire and $\bigcap_i B_{n,i} = \emptyset$ we have also that

$$U_n = \bigcup_i B^o_{n,i+1} \setminus B_{n,i}$$

is open and dense in B. Since p is minimal

$$p : p^{-1}(B^o_{n,i+1} \setminus B_{n,i}) \to B^o_{n,i+1} \setminus B_{n,i}$$

is also minimal. Hence,

$$p^{-1}(B^o_{n,i+1} \setminus B_{n,i}) \subseteq T_{n,i+1}.$$

That is, for

$$b \in B^o_{n,i+1} \setminus B_{n,i}$$

$$p^{-1}(b) \subseteq \{t : i/2^n < g(t) \leq i+1/2^n\}$$

Clearly for $b \in \bigcap_{n=1}^{\infty} U_n$, g is constant on $p^{-1}(b)$. Define, $h : B \to \mathbb{R}$ by

$$h(b) = \inf \{g(t) : t \in p^{-1}(B)\}.$$

Actually, more is true: with $p : T \to B$ as above and $g : T \to \mathbb{R}$ (or even into a separable metric space) a Borel function. Then, there exists a $G \subseteq B$, dense G_δ, such that g is constant on $p^{-1}(b)$ for each $b \in G$, and the function

$$h : G \to \mathbb{R}$$

defined by $h(b) = g(t)$, $t \in p^{-1}(b)$ is continuous on G. This is easy to see because

$$\beta = \{g : g : T \to \mathbb{R} \text{ for which such an } h \text{ exists}\}$$

contains characteristic functions of open and closed sets, is closed under sums and products and from standard facts about Baire spaces, if a sequence $\{g_n\} \subseteq \beta$ and $g_n \to g$ pointwise, then g is in β. Thus, β must contain all Borel functions.

All of this is well known.

We shall define a subset C of a topological space to be residual if for any $U \neq \emptyset$ open there exists $\emptyset \neq V$ open, $V \subseteq U$, such that $V \setminus C$ is first category.

<u>Proposition B:</u> In a Baire space B a residual set C contains a dense G_δ set; hence, the countable intersection of residual sets is residual.

<u>Proof:</u> Define

$$\mathcal{U} = \{V : V \neq \emptyset \text{ open, } V \setminus C \text{ is first category}\}$$

By hypothesis, $\cup\{V : V \in \mathcal{U}\}$ is an open dense subset of B. Suppose I is a well ordered set that indexes $\mathcal{U} : \mathcal{U} = \{V_i : i \in I\}$. If 1 is the least element of I define

$$W_1 = V_1$$

and

$$W_i = V_i \setminus \overline{\bigcup_{j<i} V_j}$$

for other $i \in I$. Let $V \neq \emptyset$ be an open subset of B, then

$$\{i : V \cap V_i \neq \emptyset\} \neq \emptyset$$

has a least element k. If $k = 1$, $V \cap V_1 \neq \emptyset$; otherwise

$$V \cap W_k = V \cap [V_k \setminus \overline{\bigcup_{j<k} V_j}] \neq \emptyset.$$

In either case,

$$V \cap [\bigcup_{i\in I} W_i] \neq \emptyset.$$

Hence, $\bigcup_i W_i$ is an open dense set. For each V_i, let $N_{i,q}$ $q = 1,2,\ldots$ be closed nowhere dense sets (which we can assume are subsets of $\overline{V_i}$) so that

$$V_i \setminus \bigcup_{q=1}^{\infty} N_{i,q} \subseteq C.$$

The set

$$G = \bigcap_{q=1}^{\infty} \bigcup_i W_i \setminus N_{i,q}$$

is a dense G_δ subset of T and $G \subseteq C$.

We have been told that this appears in [K, p. 201]. Also, it is a small piece of the proof that the Baire property sets are stable under the Souslin operation.

In proposition C below, we remark that the sets K_U need only be k-analytic in the w* topology - the proof of the proposition in this case is considerably more difficult; the utility of the approach through "fibrations" is apparent in this more general result.

<u>Proposition C:</u> ("Charting") Let X be a Banach space and $\Phi : B \to P(X^*)$ be a minimal, non-empty, compact valued usc (with respect to the weak* topology) map. Define for each $\varepsilon > 0$,

$$\mathcal{U}_\varepsilon = \{U \subseteq B,\ U \text{ open, there exists } K_U \subseteq X^*,$$

$$K_U \text{ weak}^* \text{ compact, } K_U \in \mathcal{C},\ \Phi(U) \subseteq K_U + B_{X^*}(0,\varepsilon)\}.$$

Suppose that for each $\varepsilon > 0$

$$U_\varepsilon = \cup\{U : U \in \mathcal{U}_\varepsilon\}$$

is dense in B. Then, there exists a dense G_δ subset G of B such that for each $b \in G$, $\Phi(b)$ is a singleton.

Proof: Observe, that for each $\varepsilon > 0$ and each $U \in \mathcal{U}_\varepsilon$, there exists a function

$$f : U \to X^*$$

that is continuous at each element of a dense G_δ set G_U of U, and

$$\text{dist}\,(f(b),\Phi(b)) \leqq \varepsilon$$

for all $b \in U$. The function

$$g : K_U \times \bar{B}_{X^*}(0,\varepsilon) \to X^*$$

defined by addition is perfect (with respect to its image). The multivalued map

$$g^{-1}\Phi : U \to K_U \times \bar{B}_{X^*}(0,\varepsilon)$$

is usc compact valued. There exist $\lambda : U \to K_U$, and $\xi : U \to \bar{B}_{X^*}(0,\varepsilon)$, λ continuous at each point of a dense G_δ subset G_U of U (remember Lemma a), and

$$g(\lambda(b),\xi(b)) = \lambda(b) + \xi(b) \in \Phi(b).$$

Define $f_U(b) = \lambda(b)$. Consider for each $n = 1,2,3,\dots$ the subsets $G_{U,n}$ of each non-empty $U \in \mathcal{U}_{1/n}$ and the functions $f_{U,n} : U \to x^*$ obtained as above. We know, form Proposition B, that

$$\bigcap_{n=1}^{\infty} \cup \{G_{U,n} : U \in \mathcal{U}_{1/n}\}$$

contains a dense G_δ subset G of B. Let $b_0 \in G$ and W be a weak* neighbourhood of the origin in X^*. Choose V another such neighborhood such that

$$t\bar{V} \subseteq W \quad \text{for all} \quad t, |t| \leqq 3.$$

Let $b_0 \in G$ and $x_0^* \in \Phi(b_0)$, x_0^* a weak* cluster point of

$f_{U,n}(B_o), b_o \in G_{u,n}, G_{u,n} \in U \in \mathcal{U}_{1/2}$. Choose m large enough so that

$$\bar{B}_{X^*}(0,1/m) \subseteq V.$$

There exists $q \geqq m$ such that $f_q(b_o) \in x_o^* + V$.

There exists a non-empty open set Q, $Q \subseteq U$, $b_o \in Q$, $f_q(Q) \subseteq x_o^* + V$. Thus, for $b \in Q$,

$$\Phi(b) \cap [x_o^* + 2V] \neq \emptyset.$$

Since Φ is minimal

$$\Phi(b) \subseteq x_o^* + 2\bar{V} \quad \text{for} \quad b \in Q,$$

or

$$\Phi(Q) \subseteq x_o^* + W.$$

Since W was an arbitrary neighbourhood of the origin.

$$\Phi(b_o) \subseteq \{x_o^*\}.$$

A proof of Theorem 1 can also be found in [S5].

<u>Theorem 1:</u> Let $Y \in \mathcal{S}$ and $T : X \to Y$ be an operator such that T^{**} is one to one. Then $X \in \mathcal{S}$.

<u>Proof:</u> Since T^{**} is $1 : 1$, $T^*(Y^*)$ is norm dense in X^*; that is, for each $\varepsilon > 0$,

$$X^* = \bigcup_{n=1}^{\infty} \{T^*y^* : \|y^*\| \leqq n\} + B_{X^*}(0,\varepsilon).$$

Suppose Φ is a minimal usc compact valued map from B to X^*; let U be a non-empty open subset of B. The minimality and Baire category theorem tell us that there exist m and V open and non-empty such that

$$V \subseteq U$$

and

$$\Phi(V) \subseteq \{T^*y^* : \|y^*\| \leqq m\} + B_{X^*}(0,\varepsilon).$$

Since $\{T^*y^* : \|y^*\| \leq m\} \in \mathcal{C}$ we need only apply Proposition C.

Theorem 2: Suppose X is an Asplund space, X is a subspace of the Banach space Y, and $Y/X \in \mathcal{S}$. Then, Y is in $\mathcal{S}$.

Proof: Suppose Φ is as above with values in Y^*. Let $I : X \to Y$ denote the containment operator. The map $I^*\Phi$ is also minimal. If we define

$$\mathcal{U}_\varepsilon = \{U \subseteq B : \operatorname{diam}(I^*\Phi(U)) \leq \varepsilon\}$$

then

$$U_\varepsilon = \cup\{U \in \mathcal{U}_\varepsilon\}$$

is dense in B. Choose any U non-empty in $\mathcal{U}_\varepsilon$. There exists $y_o^* \in Y^*$ such that

$$\Phi(U) \subseteq \overline{B}(y_o^*,\varepsilon) + X^\perp = (y_o^* + X^\perp) + \overline{B}(0,\varepsilon).$$

Again, by minimality and the Baire category theorem there exist m and a non-empty open set V, $V \subseteq U$ such that

$$\Phi(V) \subseteq [y_o^* + \{y^* : \|y^*\| \leq m,\ y^* \in X^\perp\}] + \overline{B}_{X^*}(0,\varepsilon).$$

Since $X^\perp$ is weak* homomorphic to $(Y/X)^*$, the conditions of Proposition C are satisfied. We have that Φ is single valued on a dense G_δ-set.

Theorem 3: Let I be an (arbitrarily large) set and $X_i \in \mathcal{S}$, $i \in I$. Then,

$$(\sum_i X_i)_{\ell_p}, \qquad 1 < p < +\infty,$$

and

$$(\sum_i X_i)_{c_o}$$

are in $\mathcal{S}$.

Proof: From Theorem 1 it suffices to prove the result for $(\sum_i X_i)_{c_o}$.

First, some notation: for each subset $J \subseteq I$ define $X_J^* = (\sum_{i \in I} X_i^*)_{\ell_1}$, if $J = I$, and $X_J^* \subseteq X_I^*$ to be those $(x_i^*) \in X_I^*$ such that $x_i^* = 0$ if $i \notin J$. Let Φ be minimal. Since the norm on X_I^* is lower semi-continuous in the weak*-topology we know there exists a dense G_δ subset of G of B such that for $b_o \in G$ (Proposition A),

(1) $\Phi(b_o)$ lies in some sphere

(2) $g(b) = \inf \{\sum_i \|x_i^*\| : (x_i^*) \in \Phi(b)\}$ is continuous at b_o.

Fix any $b_o \in G$. Since g is continuous at b_o, there exist V open and $b_o \in V$ such that

$$|g(b) - g(b_o)| < \varepsilon/2$$

for all $b \in V$. Choose $(x_i^*) \in \Phi(b_o)$ and choose F a finite subset of I and $x_i \in X_i$, $\|x_i\| = 1$, $i \in F$, such that

$$\sum_i \|x_i^*\| < \sum_{i \in F} x_i^*(x_i) + \varepsilon/2 .$$

From the upper semi-continuity, we know that

$$W = \{b \in V : (y_i^*) \in \Phi(b) \ \sum_{i \in F} y_i^*(y_i) > \sum \|x_i^*\| - \varepsilon/2\}$$

is open and non-empty.

Also, from the choice of V we know that for $b \in V$

$$g(b) < g(b_o) + \varepsilon/2.$$

That is, for $b \in V$,

$$\phi \neq \Phi(b) \cap \{(y_i^*) : \|(y_i^*)\| \leqq g(b_o) + \varepsilon/2\}$$

and we have from the minimality of Φ that

$$\Phi(V) \subseteq \{(y_i^*) : \|y_i^*\| \leqq g(b_o) + \varepsilon/2\}$$

We need to show that

$$\Phi(W) \subseteq X_F^* + \bar{B}_{X_I^*}(0,\varepsilon).$$

For $b \in W$, $(y_i^*) \in \Phi(b)$

$$\left|\sum_i \|y_i^*\| - \sum_i \|x_i^*\|\right| < \varepsilon/2$$

and

$$-\varepsilon/2 + \sum_i \|x_i^*\| < \sum_{i \in F} y_i^*(x_i) .$$

Therefore,

$$\varepsilon \geqq \sum_i \|y_i^*\| - \sum_{i \in F} y_i^*(x_i) = \sum_{i \notin F} \|y_i^*\| + \sum_{i \in F} (\|y_i^*\| - y_i^*(x_i)).$$

For $i \in F$, $\|y_i^*\| - y_i^*(x_i) \geqq 0$. Thus, $\sum_{i \notin F} \|y_i^*\| \leqq \varepsilon$. We have that

$$(y_i^*) \in X_F^* + \bar{B}_{X_I^*}(0,\varepsilon).$$

We have shown that

$\mathcal{U}_\varepsilon = \{U : U$ open $\subseteq B$, $\exists$ F finite $\exists$ m a positive integer such that

$$\Phi(U) \subseteq \{(x_i^*) \in X_F^* : \|(x_i^*)\| \leqq m\} + \bar{B}_{X_F^*}(0,\varepsilon)\}\}$$

has dense union in B. Since each bounded subset of each X_F^*, F finite, is in C we know that $X_I \in S$.

Theorem 4 below is only another variation of Theorem 1-3. Theorem 5 requires the non-trivial result of G. Debs [D1] that weakly k-analytic Banach spaces are in S (see also [S6]).

Theorem 4: Suppose $X_i \in S$ and

$$T : (\sum_i X_i)_1 \to Y$$

has dense range and

$$T^*(Y^*) \subseteq (\sum_i X_i^*)_{c_o}$$

Then, $Y \in \mathcal{S}$.

Theorem 5: Let (X_i) be a family of weakly k-analytic Banach spaces, then

$$(\sum_i X_i)_{c_o} \in \mathcal{S}$$

Without proofs, we list a few interesting habits of some of the inhabitants of $\mathcal{S}$.

(1) Let X be a separable Banach space, then X^* is separable if and only if X^* is weakly k-analytic (there are many proofs of this e.g. [T1] - a proof can also be squeezed out of [S1].

(2) There exists a Banach space X such that X is separable, X^* is non-separable but X^* is an Asplund space; hence $\mathcal{S}$ is strictly larger than the weakly k-analytic Banach spaces;

(3) The class of WCG (weakly compactly generated) Banach spaces is not hereditary (see [T1] and its references) but WCG spaces are stable under arbitrarily large ℓ_p sums, $1 < p < +\infty$; WCG Banach spaces are in $\mathcal{S}$;

(4) The weakly k-analytic Banach spaces are hereditary but it is not clear if they are stable under uncountable ℓ_p sums, $1 < p < +\infty$.

Example: "the top and bottom line of the dictionary square"

Let $[0,1]$ be the unit interval with the usual topology, let $S = [0,1] \times [0,1]$ be the square with the order topology given by the dictionary ordering and let $T = \{(t,0), (s,1) : 0 \leqq t, s \leqq 1\}$ be in the top and bottom line of the square. In the order topology T is a compact space. The mapping $p : T \to [0,1]$, $p(t,0) = p(t,1) = t$ is continuous and onto. Since the spaces involved are compact, p is perfect. It is easy to check that if we delete $(0,0)$ and $(1,1)$ it is also minimal perfect.

Thus, $T \notin C$ because $p^{-1}(t)$ has two points for every $0 < t < 1$. Let $C(T)$ be the continuous functions on T, $A : C[0,1] \to C(T)$ be defined by $Ag = g \circ p$ and

$$Q : C(T) \to C_o[0,1]$$

be defined by $Q(f) = (f(t,1) - f(t,0))$.

It is easy to check that Q is well defined, continuous, onto and that the kernel of Q is $A(C[0,1])$. We have that

$$C[0,1] \text{ and } c_o[0,1] \text{ are in } \mathcal{S}$$

but

$$C(T) \text{ is not in } \mathcal{S}.$$

Thus, $\mathcal{S}$ does not have the "three space property". The pathologies of $C(T)$ have been investigated in greater detail by Talagrand ([T1],[T2]).

REFERENCES

[D] G. Debs, Gateaux derivatibilité des fonctions convexes continues sur les espaces de Banach non separables, to appear.

[K] J. L. Kelley, General Topology, Van Nostrand, New York, 1955.

[S1] C. Stegall, The Radon-Nikodym property in conjugate Banach spaces, Trans. of the Amer. Math. Soc., Vol. 206, 1975.

[S2] C. Stegall, The duality between Asplund spaces and spaces with the Radon-Nikodym property, Israel J. of Math., Vol. 29, No. 4, 1978, 408-412.

[S3] C. Stegall, The Radon-Nikodym property in conjugate Banach spaces, II, Trans. of the Amer. Math. Soc., Vol. 264, No. 2, April 1981, 507-519.

[S4] C. Stegall, A class of topological spaces and differentiation of functions on Banach spaces, Vorlesungen aus dem Fachbereich Mathematik der Universität Essen, Heft 10, 1983, 63-77.

[S5] C. Stegall, Gateaux differentiation of functions on a certain class of Banach spaces, Func. Analysis: Surveys and Recent Results III, North Holland, 1984, 35-46.

[S6] C. Stegall, to appear.

[T1] M. Talagrand, Espaces de Banach faiblement k-analytiques, Annals of Math. 110, (1979), 407-438.

[T2] M. Talagrand, Renormages de quelques C(K), to appear.